RD
559
J63

THE
HAND
BOOK

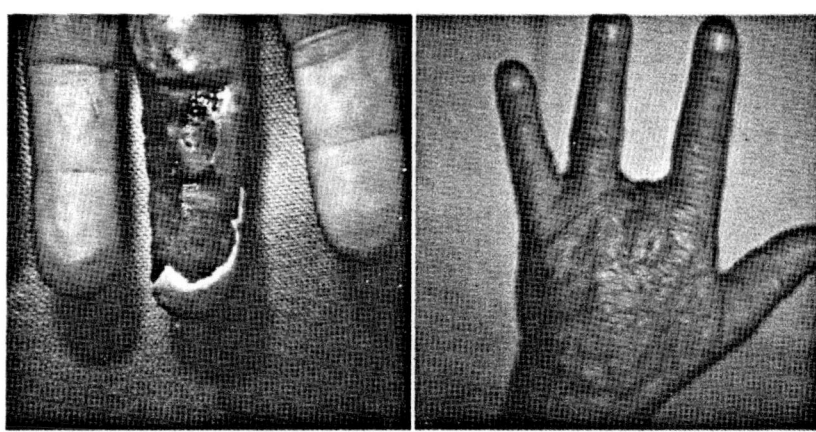

New industrial techniques create new types of injuries. Particularly deceptive is the injury sustained when paint, diesel oil, plastic, or grease is accidentally injected into a hand or finger by high-pressure injection equipment. Because the wound of entrance is minute and the initial discomfort minimal, the injury is often dismissed as trivial. However, formal surgical evacuation of the injected material on the day of injury is essential. The finger shown on the left was inadequately treated following a paint injection injury. Skin necrosis and infection made amputation necessary, leaving the patient as seen on the right.

THE HAND BOOK

By

MOULTON K. JOHNSON, M.D.

*Chief, Hand Clinic
Assistant Clinical Professor of Surgery/Orthopedics
U.C.L.A. School of Medicine
Los Angeles, California
Senior Surgeon, Santa Monica Hospital
Senior Surgeon, St. John's Hospital
Santa Monica, California*

CHARLES C THOMAS • PUBLISHER
Springfield • Illinois • U.S.A.

Published and Distributed Throughout the World by
CHARLES C THOMAS • PUBLISHER
BANNERSTONE HOUSE
301-327 East Lawrence Avenue, Springfield, Illinois, U.S.A.

This book is protected by copyright. No part of it may be reproduced in any manner without written permission from the publisher.

© 1973, *by* CHARLES C THOMAS • PUBLISHER
ISBN 0-398-02595-9
Library of Congress Catalog Card Number: 72-81702

With THOMAS BOOKS *careful attention is given to all details of manufacturing and design. It is the Publisher's desire to present books that are satisfactory as to their physical qualities and artistic possibilities and appropriate for their particular use.* THOMAS BOOKS *will be true to those laws of quality that assure a good name and good will.*

Printed in the United States of America
W-2

TO

ROBERT M. McCORMACK, M.D.,

who introduced me to the art of hand surgery, and not only encouraged my interest but also indoctrinated me in its basic principles. He invested more time and patience in that task than anyone could expect, even from a devoted older brother.

INTRODUCTION

The Hand Book is written for the family physician and other nonsurgeons who are frequently the first to see the patient with an acute injury or other problems involving the hand. The primary aim is to present a practical guide for the rapid evaluation of hand problems, rather than provide "how-to-do-it" instructions for treatment. It is not intended to be original and draws freely from the literature.

The hand is both compact and complex, containing 27 bones and 18 intrinsic muscles. In addition, the tendons of 21 extrinsic muscles insert into the hand. The total number of possible combinations of injury is incalculable. Even quite small wounds or relatively undramatic blunt trauma are capable of producing serious problems. *The Hand Book*, therefore, emphasizes those conditions or injuries which may appear to be minor but which can actually lead to permanent disability if not recognized and treated early. The importance of preservation of function is also stressed, since the function of the hand cannot be reproduced by even the best of mechanical devices.

The Hand Book also attempts to give the practitioner some knowledge of what can be accomplished in a given situation by a specialist in hand surgery, as well as some insight into the limits of his art.

ACKNOWLEDGMENTS

I WOULD LIKE TO express my appreciation to the people who helped me prepare this monograph. Edna and Eric contributed valuable editorial assistance. Bud and Ed coached me in photography. Cass, Lynda, Phil, and Sherita, along with many patients, posed for the illustrations, and Ginny did not begrudge the many evenings spent in the darkroom.

M.K.J.

CONTENTS

	Page
Introduction	vii
Acknowledgments	ix

Chapter
- I. NOMENCLATURE 3
- II. ACUTE INJURIES 7
- III. INFECTIONS 62
- IV. DEGENERATIVE DISEASES 67
- V. RHEUMATOID ARTHRITIS 84
- VI. DUPUYTREN'S CONTRACTURES 95
- VII. TUMORS AND MASSES 99

Index 109

THE HAND BOOK

Chapter I

NOMENCLATURE

THERE ARE SEVERAL systems of nomenclature in current use to describe the parts of the hand. In order to avoid confusion, the system used here is outlined below.

The term "hand" will include not only the carpal and metacarpal areas but the thumb and four fingers as well. The hand is considered to have three surfaces, viz. dorsal—the back of the hand, volar—the front or palmar surface of the hand, and radial (Figure 1)—the dorsal surface of the first (thumb) metacarpal and the radial side of the second (index) metacarpal.

The four fingers are index, long, ring, and little fingers. The three joints of each finger are referred to as MP (metacarpophalangeal), PIP (proximal interphalangeal), and DIP (distal interphalangeal). (See Figs. 2 and 3.) The two joints of the thumb are referred to as MP (metacarpophalangeal) and IP (interphalangeal). Extrinsic muscles are defined as those which arise in the forearm and insert within the hand, and intrinsic muscles those which both take origin and insert within the hand.

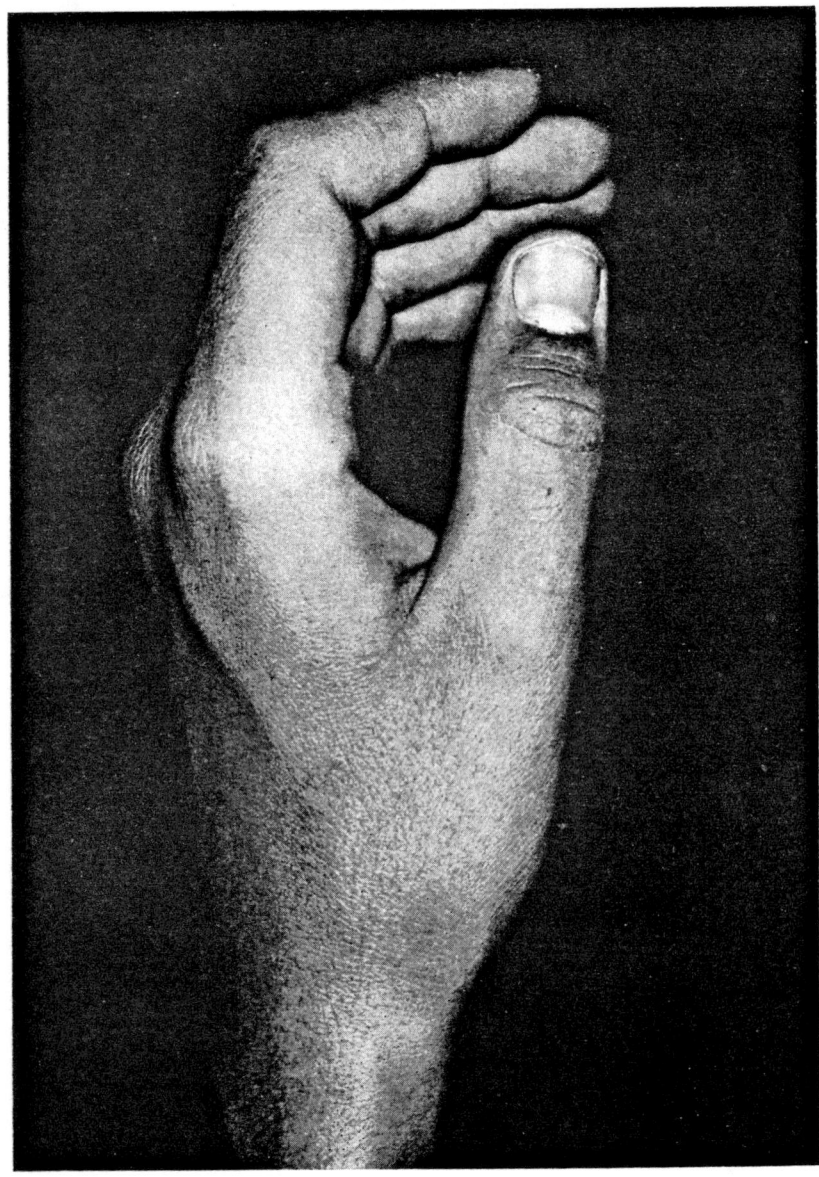

Figure 1. The radial side of the hand.

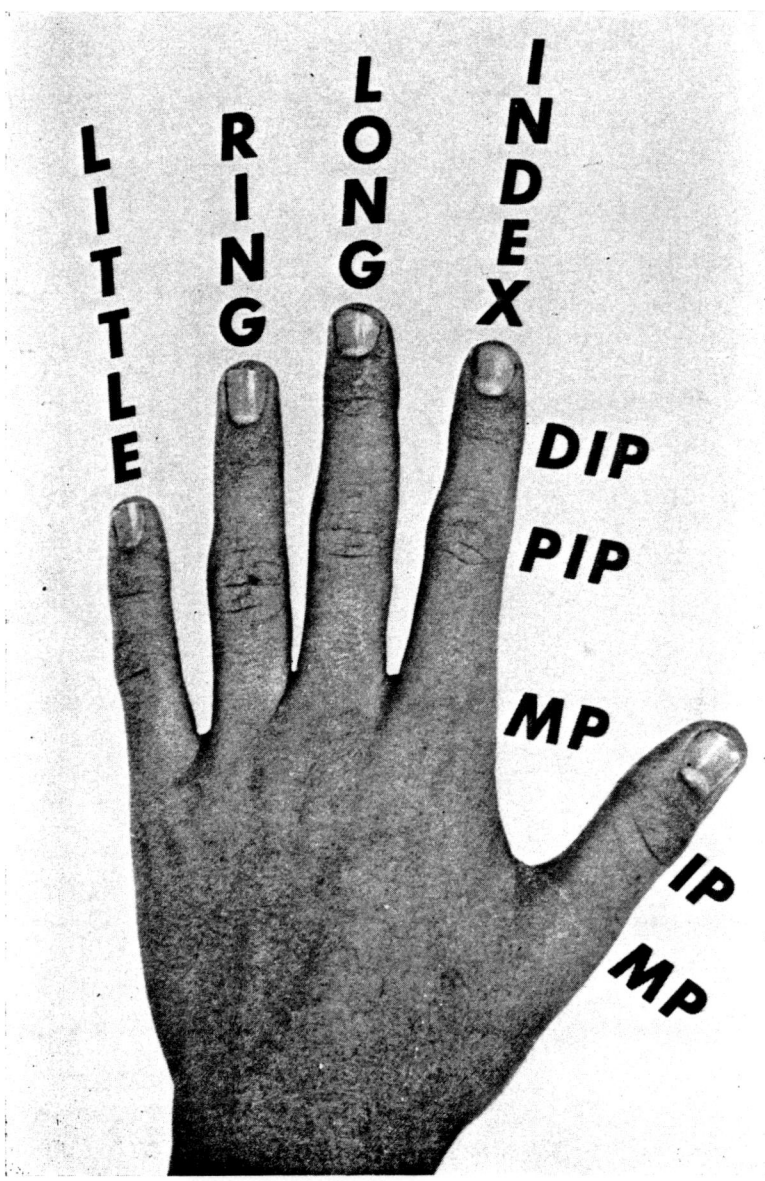

Figure 2. The dorsal surface of the hand.

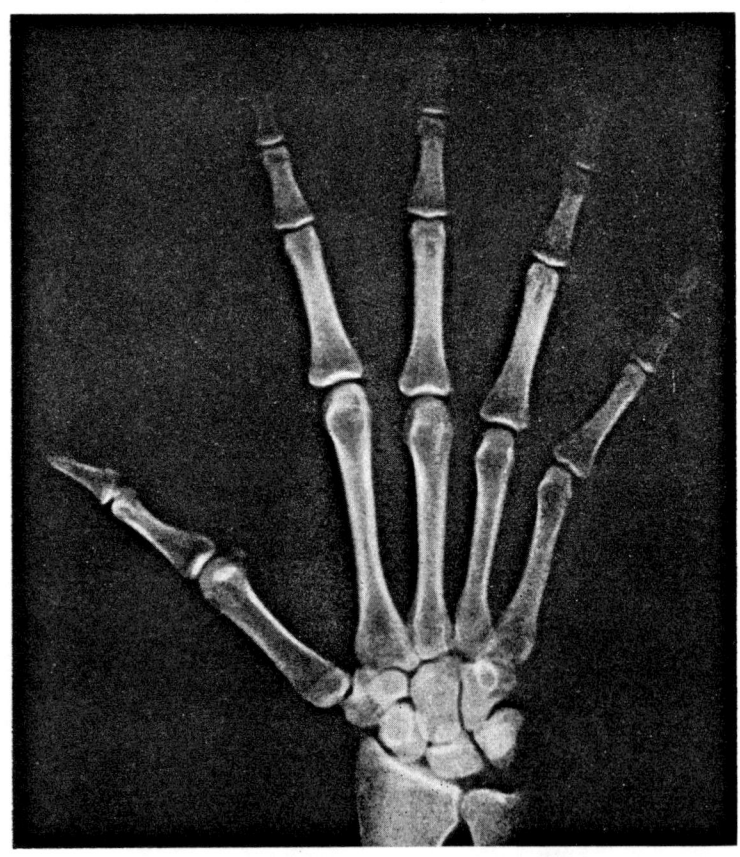

Figure 3. Anteroposterior x-ray view of the hand.

Chapter II

ACUTE INJURIES

LACERATIONS

THE ESSENTIAL ELEMENT in the diagnosis of lacerations about the hand is applied anatomy.

The deep structures on the volar side of the hand and wrist are more essential to function than those on the dorsal side. Volar lacerations are, therefore, potentially more serious than dorsal lacerations.

There are four consecutive steps that should be followed in order to properly evaluate and diagnose the nature and degree of loss of function:

1. Inspect the external wound carefully and consider which underlying structures might have been injured.

2. Test the function of each nerve and muscle-tendon unit distal to the wound.

3. If necessary, use local anesthetics to carefully inspect the depths of the wound whenever a tendon might have been partially divided. Partial severance of up to 95 percent of the tendon is easily overlooked, since a partially severed tendon can move the part through a full range of motion on examination and yet rupture later when full force is applied. *Always* evaluate nerve function *before* using local anesthesia.

4. If the injury might have produced a fracture or dislocation, obtain appropriate x-rays.

Median and Ulnar Nerves

Complete severance of the median or ulnar nerve in an adult can be a catastrophe, as the patient is disabled during the many months required for regeneration of the nerve. Since the results of nerve repair are often disappointing, additional operations

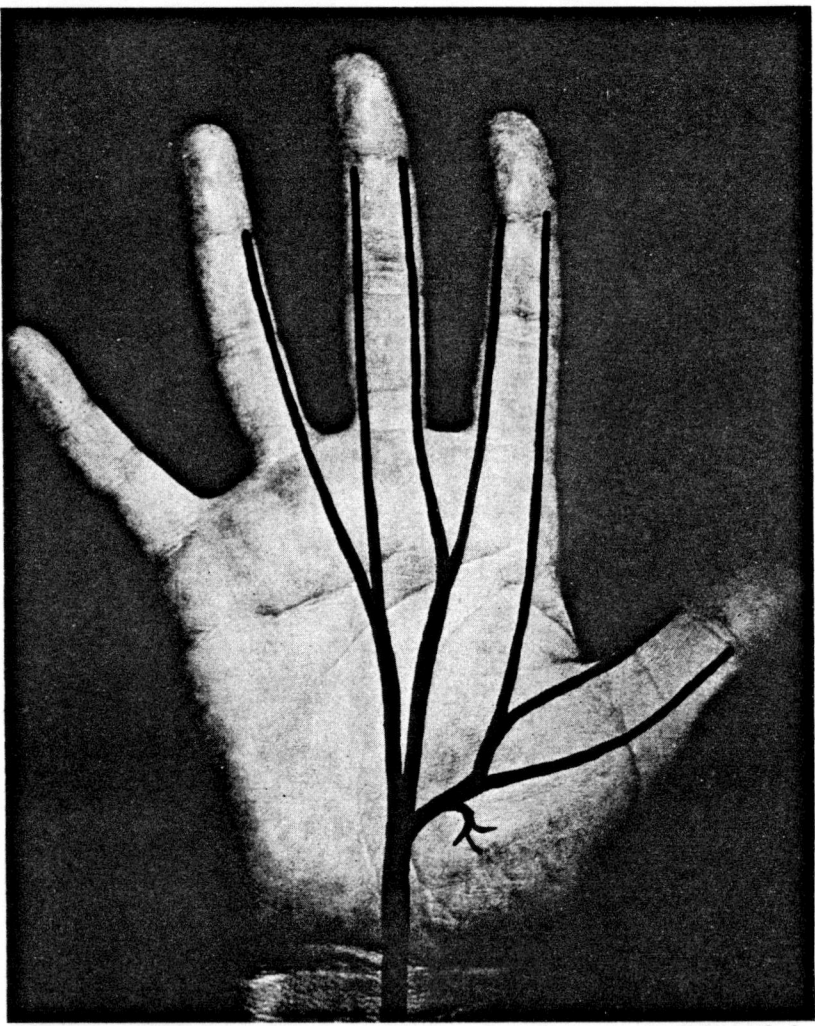

Figure 4. Schematic drawing of the course of the main branches of the median nerve. Repair of the digital sensory branches is not feasible at the distal phalanges.

such as tendon transfers designed to substitute for lost function, may lead to further months of disability. Generally speaking, the older the patient, the worse the outcome.

Severance of the median nerve at the wrist results in inability to oppose the thumb, and in anesthesia of the volar surface of the thumb, index, long, and the radial half of the ring finger, with sensation being the most important function lost. The course of the median nerve is shown schematically in Figure 4. Recovery of sensation is rarely complete following repair of the median nerve in the adult. Although a casual examination may show that the patient appreciates pin-prick and light touch in a normal manner, a careful examination reveals loss or diminution of two-point discrimination. The patient thereby loses a measure of tactile gnosis, which is essential for rapid and precise manipulation of small objects and for the tactile differentiation of objects, e.g. coin or key in a pocket. If thumb opposition does not return, this function may be restored satisfactorily by appropriate tendon transfer.

Severance of the ulnar nerve at the wrist results in the loss of function of the dorsal and volar interossei, the adductor pollicis, and the hypothenar muscles, as well as the lumbricals to the ring and little fingers. The patient exhibits anesthesia over the volar surface of the little finger and ulnar half of the volar surface of the ring finger. The course of the ulnar nerve is shown schematically in Figure 5. The key function lost is the use of the adductor pollicis and the first dorsal interosseous. The adductor is essential to strong pinch by the thumb, and the first dorsal interosseous stabilizes the index against the thumb during pinch. Strength of the ulnar-innervated intrinsic muscles fails to return to functional levels in 75 percent of all ulnar nerve repairs in the adult patient. Although appropriate tendon transfers are of help, the hand will usually remain clumsy and relatively weak.

Testing the function of the median and ulnar nerves is the important first step in the evaluation of open injuries about the palm and volar aspect of the wrist. First, assume that there might have been injury to a nerve. Examine sensation of each digit. Next, have the patient oppose the tip of the thumb to the tip of the little finger to evaluate the median nerve supply to the thenar intrinsic muscles. When the median nerve or its motor branch has been severed, the patient can flex and adduct the thumb, but cannot rotate the thumb out from the palm and oppose the tip of the little finger (Fig. 6). An occasional patient will retain weak opposition when his ulnar nerve participates in the innervation of one or more thenar muscles. Last, examine the

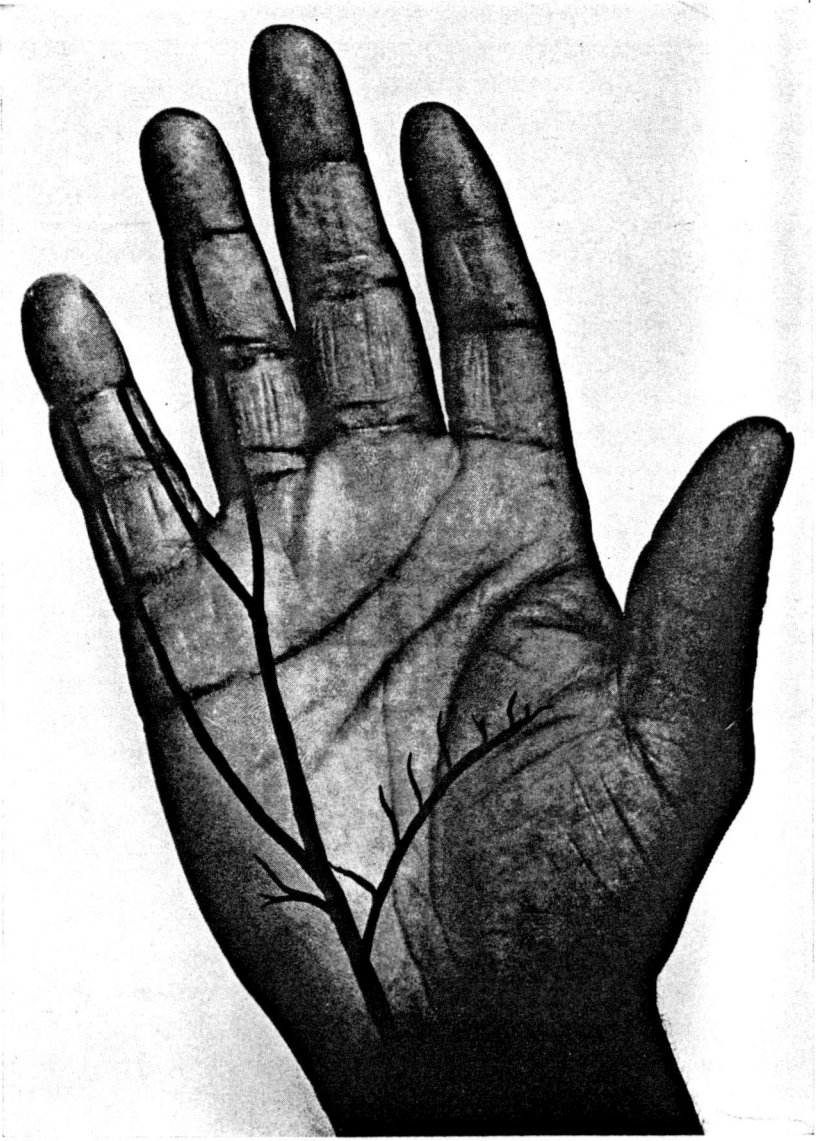

Figure 5. The course of the main branches of the ulnar nerve drawn schematically.

intrinsic muscles supplied by the ulnar nerve. Test wide abduction of the fully extended fingers (doral interossei) as seen in Figure 7, and the adduction of the fully extended fingers and

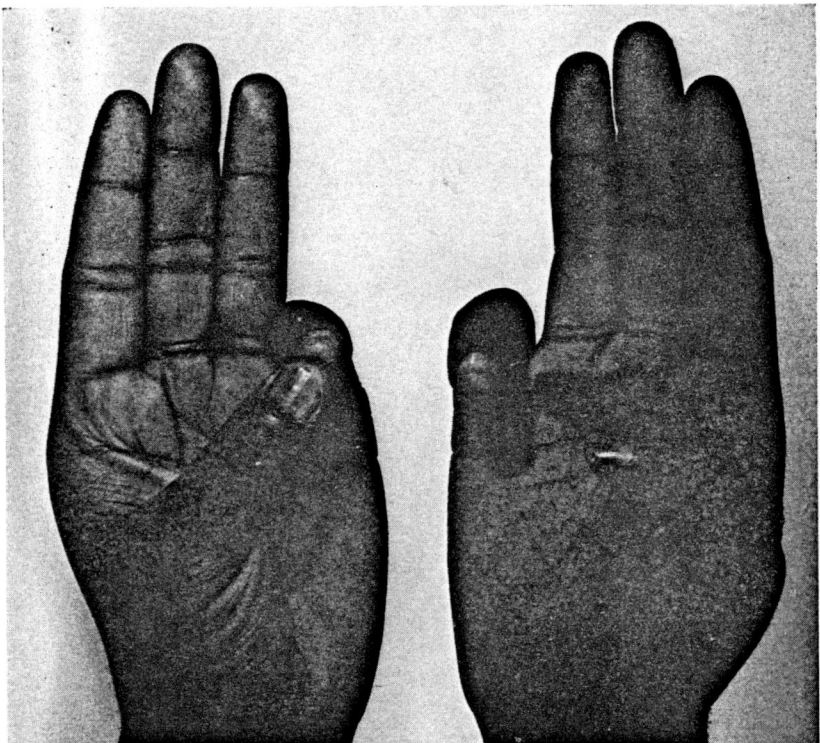

Figure 6. The median motor branch has been severed in the right hand. Although the patient can flex and adduct the thumb, he is unable to oppose it to the tip of the little finger. The left hand is normal.

thumb (volar interossei and adductor pollicis) as in Figure 8. Prompt diagnosis is the key factor in obtaining the best possible end result. Most surgeons advocate secondary repair of major mixed motor and sensory nerves. (Sensory nerve branches in the hand and digits can be repaired primarily if the wound is clean.) Nevertheless, the surgeon should approximate the two ends of the nerve on the day of injury to prevent retraction. Secondary repair is much easier if the surgeon does not have to overcome a sizeable gap.

Extensor Tendons

Repairs of the extensor tendons usually give satisfactory results. Even when a severed extensor tendon is overlooked and

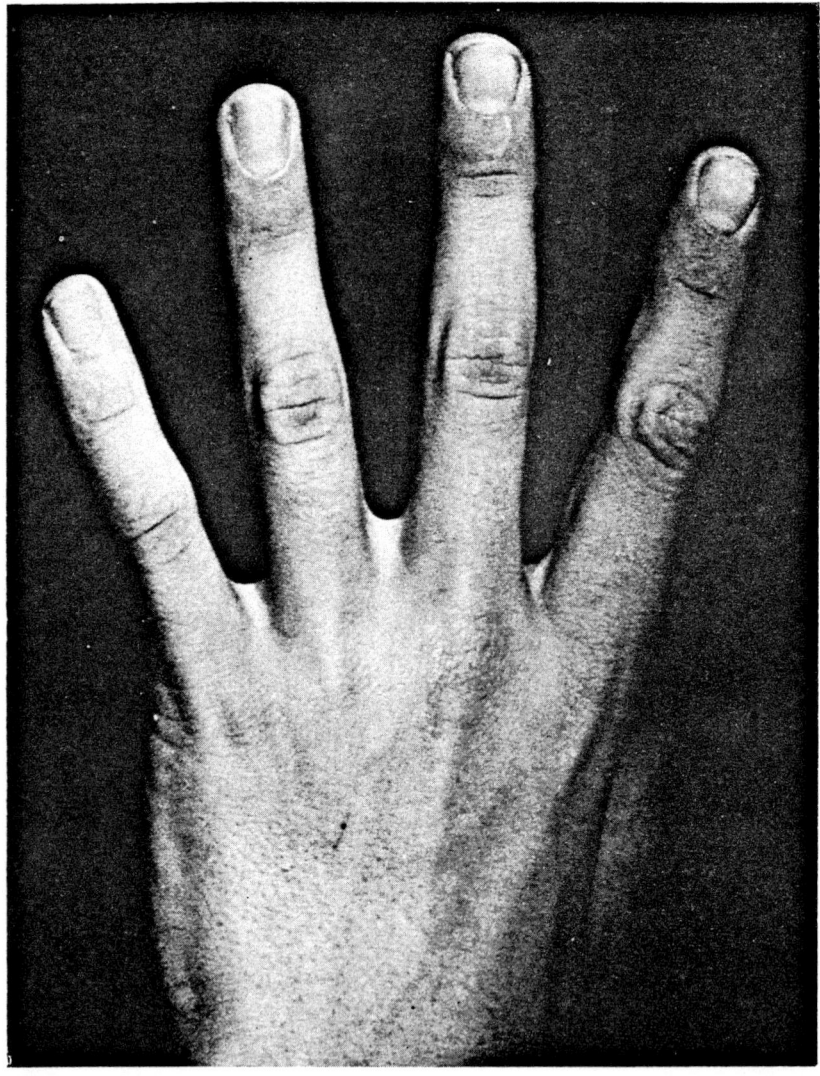

Figure 7. To test the dorsal interossei, the patient is asked to fan the fully extended fingers into wide abduction.

the skin laceration sutured, secondary repair of the tendon after the wound has healed is frequently successful.

Lacerations on the radial side of the wrist may divide the radial artery and injure sensory branches of the radial nerve.

Acute Injuries 13

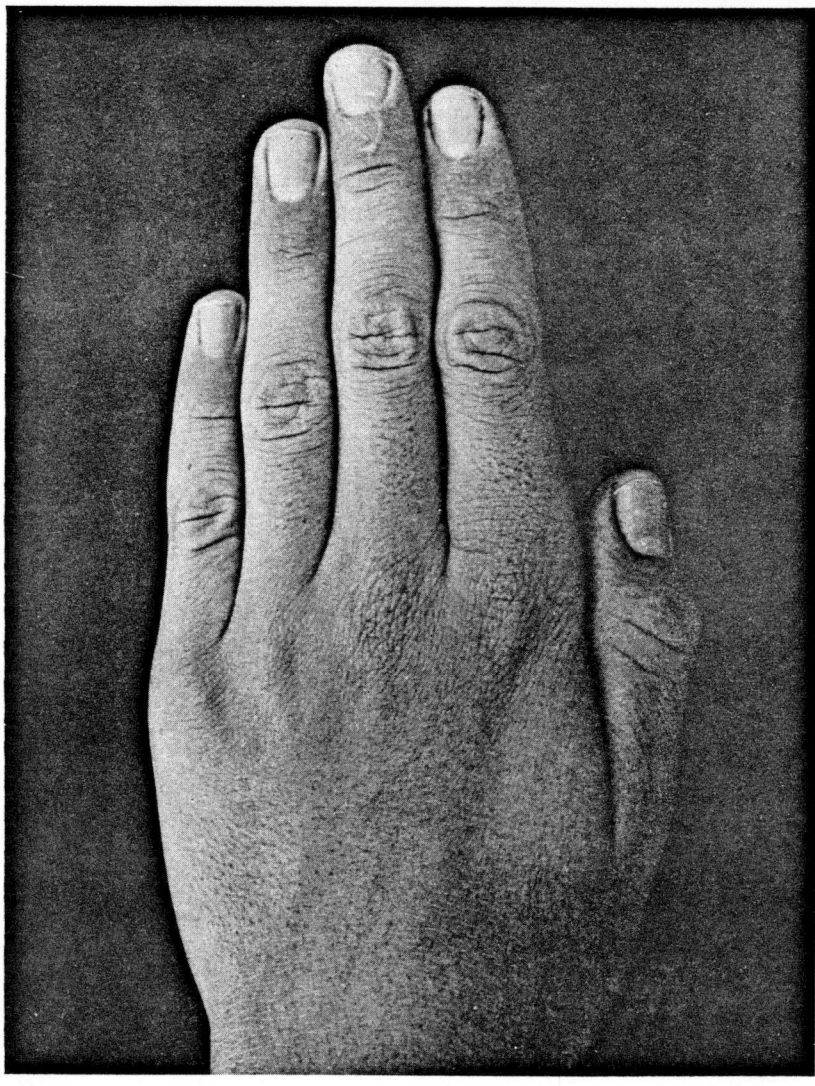

Figure 8. To test the volar interossei, the patient is asked to adduct the fully extended fingers.

However, the sensory loss is usually inconsequential, although occasionally a painful neuroma will result. If the patient has a normal ulnar artery (as evidenced by a good ulnar pulse and no loss of normal skin color in the digits), the radial artery may

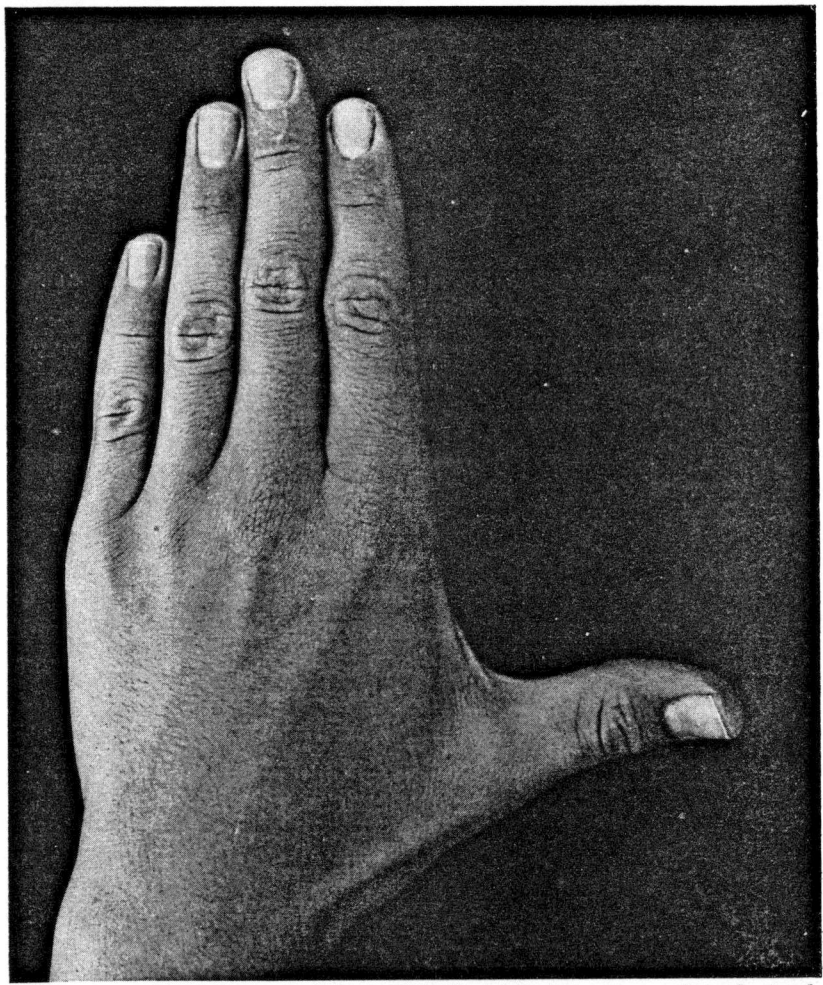

Figure 9. The ability to abduct fully the thumb metacarpal and simultaneously extend the MP and IP joints of the thumb demonstrates the integrity of the three tendons, extensor pollicis longus and brevis, and abductor pollicis longus.

simply be ligated rather than repaired. Injury to the extensor pollicis longus and brevis and/or to the abductor pollicis longus is the major concern, and superficial examination may mislead the physician. To rule out severance of these tendons, the patient must be able to fully extend both joints of the thumb and simultaneously abduct the thumb (Fig. 9). The best evaluation is a side-to-side comparison with the uninjured hand.

The wrist extensors, finger extensors, and the proximal portion of the extensor pollicis longus tendon are vulnerable to injury on the dorsum of the wrist. Severance of a single wrist extensor tendon is rare; in such cases the severed tendon end is the best evidence of such an isolated injury. If the wrist deviates toward the radius as the patient actively dorsiflexes the wrist, the extensor carpi ulnaris has been severed. If the dorsiflexed wrist deviates toward the ulna, it is the extensor carpi radialis longus (and perhaps the brevis as well) which has been severed.

To obtain full simultaneous extension of all finger joints, combined contraction of both the extensor digiti communis and at least one interosseous or lumbrical muscle per finger is necessary. The function of the extensor digiti communis is demonstrated, therefore, by having the patient simultaneously extend all three joints of all four fingers. To demonstrate function of the extensor indicis proprius, ask the patient to point with the extended index finger while flexing the other fingers into a fist (Fig. 10).

Flexor Tendons

There are three sets of muscles which participate in flexion of the four fingers. The profundus muscles, one for each of the four fingers, arise in the forearm and insert into the distal phalanx. Because they cross each of the three joints of each finger (MP, PIP, DIP), they produce flexion at each joint. The four sublimis muscles have shorter tendons which insert into the middle phalanx of each finger. The sublimi flex the MP and PIP joints, but of course have no effect on the DIP joints. The dorsal and volar interossei and the lumbricals are flexors of the MP joints of the four fingers, in addition to their previously described roles in extension of PIP and DIP joints and abduction and adduction of the fingers. The flexor pollicis longus is, in effect, the profundus of the thumb, producing flexion at the MP and IP joints of the thumb. There is no equivalent of a sublimis for the thumb, as it has only the one interphalangeal joint. The flexor pollicis brevis is a flexor of the thumb MP joint.

Because the profundi and sublimi and the flexor pollicis longus arise in the forearm and insert into the middle or distal phalanx

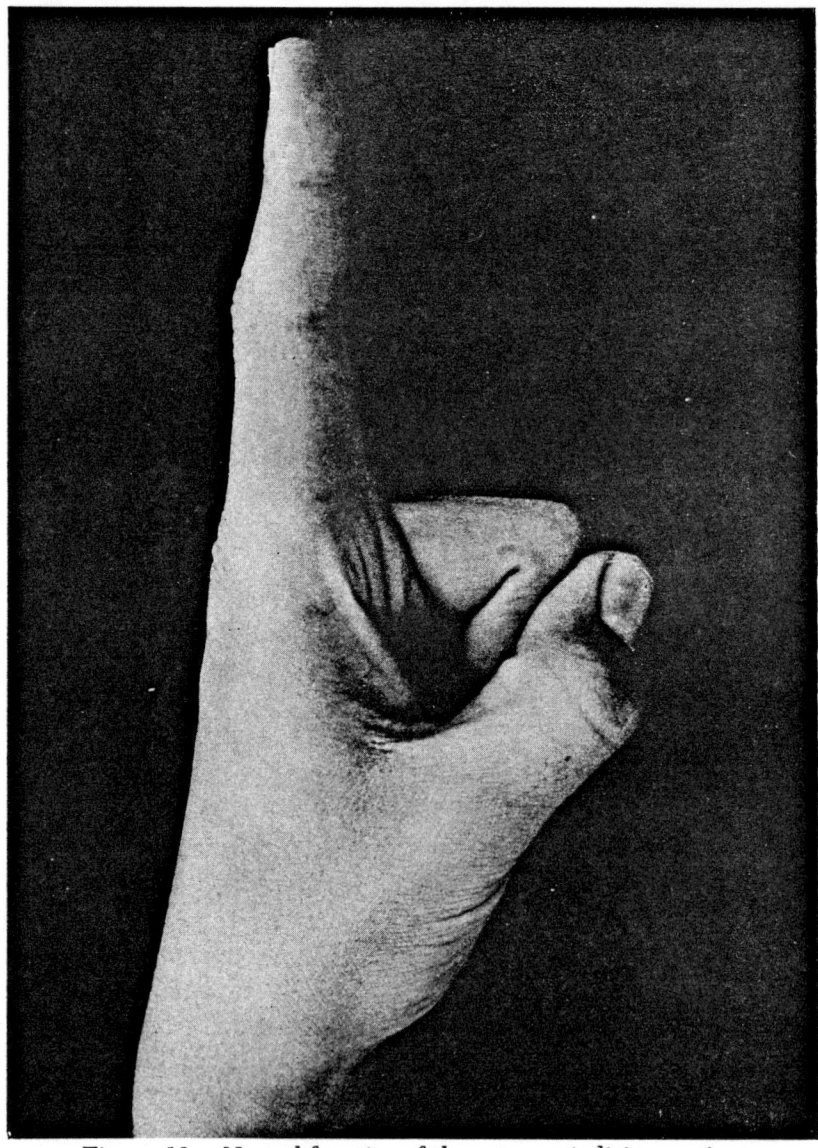

Figure 10. Normal function of the extensor indicis proprius.

of the appropriate digit, they may be severed at the wrist, in the palm, or out in the finger or thumb.

To evaluate the function of the flexor tendons of the four fingers, first have the patient make a fist. Look for active flexion at each DIP joint (Fig. 11). If the DIP joints all flex actively,

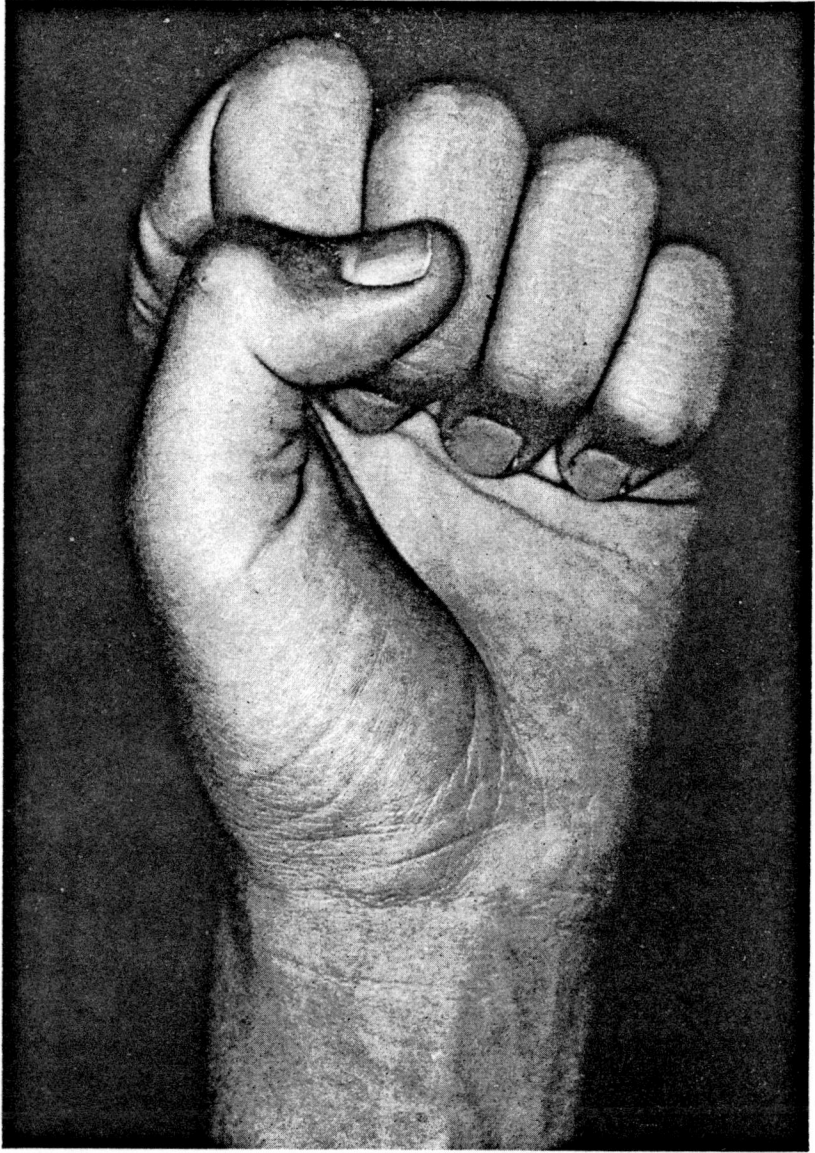

Figure 11. If the profundus tendons are intact or only incompletely severed, the patient can make a fist.

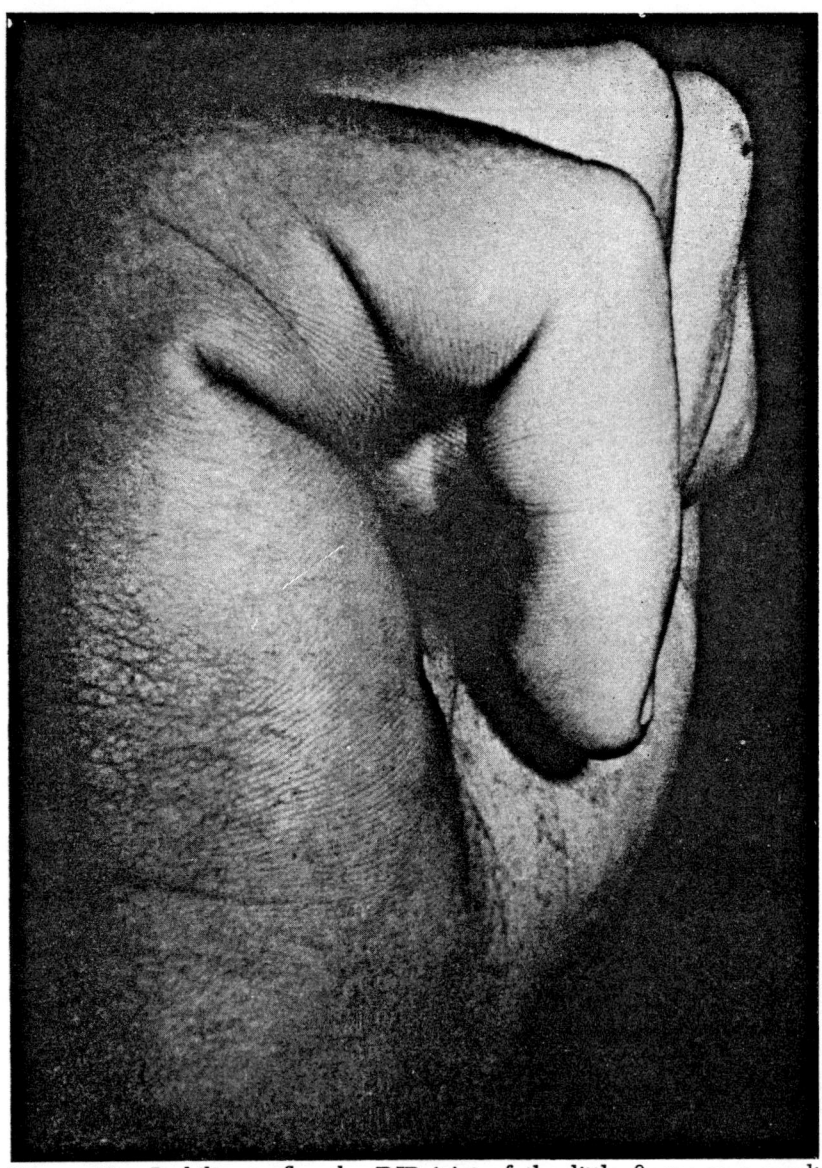

Figure 12. Inability to flex the DIP joint of the little finger as a result of laceration of the little finger profundus.

the profundi cannot have been completely severed, whereas failure to move the joint demonstrates loss of function. Figure 12 shows a patient whose little finger profundus tendon has been divided. The patient is able to flex the MP and PIP joints of the finger, but the DIP joint remains in extension as the patient tries to make a fist. Had the sublimis been severed in addition to the profundus, the patient would not have been able to flex actively either the PIP or the DIP joint. However, he would have retained the ability to flex the MP joint, using the intrinsic muscles.

Through most of its course, each sublimis tendon lies superficial to the accompanying profundus. Therefore, a superficial laceration may sever the sublimis tendon but leave the profundus tendon intact, and the patient will still be able to flex all three of his finger joints with his intact profundus. Therefore, to demonstrate severance of the sublimis alone, it is necessary to neutralize the profundus. The examiner can do this by holding the other three fingers in full extension. Figure 13 shows the examiner testing the ring finger sublimis. He holds the other fingers in full extension while asking the patient to flex the ring finger as strongly as possible. The finger is flexing at the MP and PIP joints, but not at the DIP joint, demonstrating that the sublimis tendon is intact. Had the sublimis tendon been cut, the patient would be able to flex only the MP joint. Some individuals are unable to flex the PIP joint of the little finger when the other fingers are held in extension, producing the appearance of a severed sublimis. In such a case, inspection of the wound under local anesthesia will be necessary to determine the status of the sublimis.

Evaluation of the flexor pollicis longus is straightforward. If the patient is able to actively flex the IP joint of the thumb when he makes a fist, the tendon has not been completely divided.

The three superficial tendons on the volar side of the wrist are the flexor carpi ulnaris, palmaris longus (absent in 12 percent of patients), and flexor carpi radialis. They overlie the ulnar nerve and artery, the median nerve, and the sublimis tendons. Because all three of these muscles flex the wrist, and because they are assisted in this function by the profundi and sublimi, functional testing of these muscles is difficult. Their status is best evaluated by direct inspection of the wound. Repair of a single wrist flexor or the palmaris longus is unnecessary because

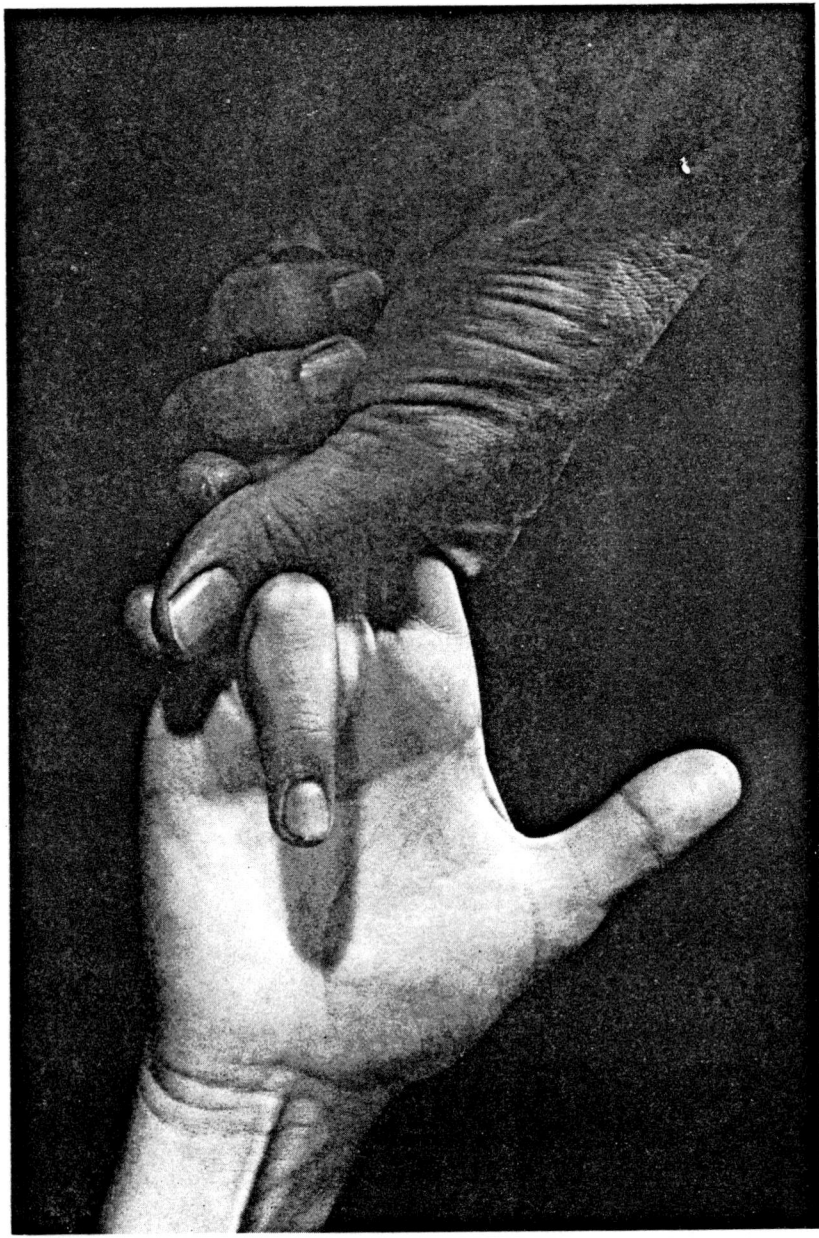

Figure 13. Test of the ring finger sublimis function. The ring sublimis is functional in this patient.

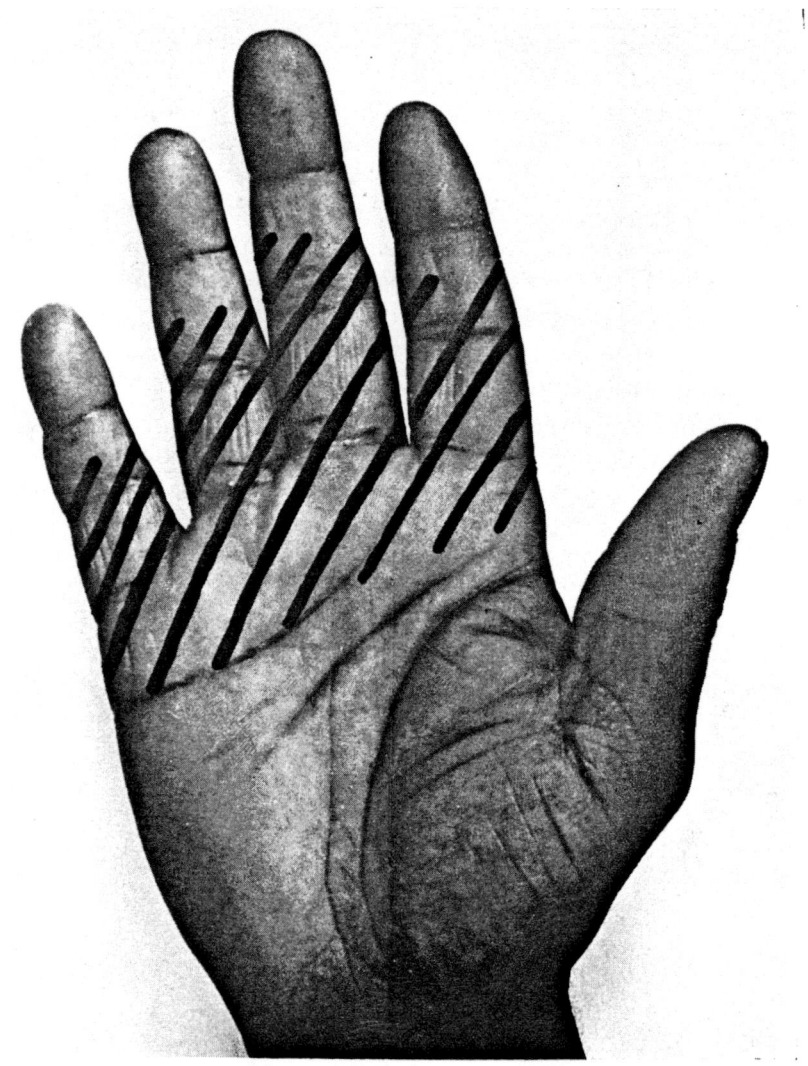

Figure 14. The lined area is known as "no-man's land."

there are so many muscles which act as wrist flexors and the loss of a single wrist flexor produces no noticeable effect.

The area extending from the distal palmar flexion crease to the middle of the middle phalanx on the volar side of each finger is known as "no-man's land" (Fig. 14). The profundus and sublimis tendons pass through this area in snug osseofibrous tunnels. The results of tendons repairs in this area in adult patients are notoriously poor, regardless of whether the surgeon employs primary or secondary repair or a free-tendon graft. Even with clean incised lacerations, the tendons often become adherent to one another and bound down to the surrounding tissues, severely limiting motion of the finger joints. In crushing injuries or when the wound becomes infected, the prognosis is even worse.

The flexor pollicis longus passes through a similar osseofibrous tunnel in its course from the distal palm to the distal phalanx, and poor results are also frequently obtained from repairs in this area.

Although tendon lacerations in the palm and at the wrist are somewhat easier to manage, it cannot be overemphasized that only a small fraction of flexor tendon repairs in adults yields completely normal function.

AMPUTATIONS

When the physician encounters a patient with a fresh traumatic amputation of the thumb or the finger, or a patient who has suffered damage which makes amputation unavoidable, he should bear in mind the primary rule of "save length." It is particularly important to preserve maximum length of the thumb since it is responsible for 40 percent of hand function. It is both unnecessary and archaic in most cases to shorten the bone in order to close the skin. Many methods are available for repair of soft tissues and closing the skin without sacrificing bone length, the most commonly employed techniques being free skin grafts and local skin flaps.

It is, of course, frequently impossible to predict whether or not an injured digit will recover sufficiently to be useful, but even in borderline cases the surgeon should attempt to salvage the digit. In some instances of severe damage in which the finger was saved, amputation may later become necessary because of pain, deficient sensation, stiffness, poor circulation, cosmetic reasons, or a combination of these factors. However, delayed elective amputation will be based on indications which are clear-cut to both the patient and the surgeon.

A patient will use a thumb stump no matter how short, simply because he has no substitute available. In the case of finger amputations, however, usefulness declines in direct proportion to decreasing stump length, although degree of usefulness varies with the patient. A finger stump of moderate length is useful and cosmetically acceptable to one patient, whereas a stump of the same length may be useless and/or cosmetically unacceptable to another. Amputation through the distal phalanx does not shorten the digit enough to reduce function significantly, nor does it change the general contour of the hand to the extent of being conspicuous. An amputation through either the DIP joint or the middle phalanx leaves a stump which is useless for the precise functions of the hand (Fig. 15) but is still useful for gross grasp (Fig. 16).

When the amputation is at or proximal to the PIP joint, the stump can be of limited use only to the person who performs the heaviest kind of manual labor. In this instance the contour of the hand is grossly altered and the cosmetic defect is obvious. Also, if the long or ring finger has been amputated at or near the MP joint, the patient constantly drops small objects through the resulting gap. Therefore, most patients with MP disarticulations or amputations of the finger proximal to the DIP joint are best served by amputation of the entire stump and most of the metacarpal. This procedure, called a *ray resection*, restores a near normal contour to the hand, and the loss of breadth of palm is of no consequence in most occupations. However, if several fingers have been amputated, a more conservative approach is indicated. Similarly, if the amputations were done because of vascular disease or other conditions in which other fingers may become involved at a later date, one should conserve all possible length.

Incomplete amputation of the distal one-fourth or one-half inch of a digit is a common injury produced by closing a door on the finger. This injury may appear hopeless because the tip is connected to the finger only by a narrow bridge of skin and subcutaneous tissue. However, the fingertip is very vascular, and more often than not the tip will survive if sutured back in place. This is especially true in small children. Furthermore, do not assume that the tip is avascular if it appears black a few

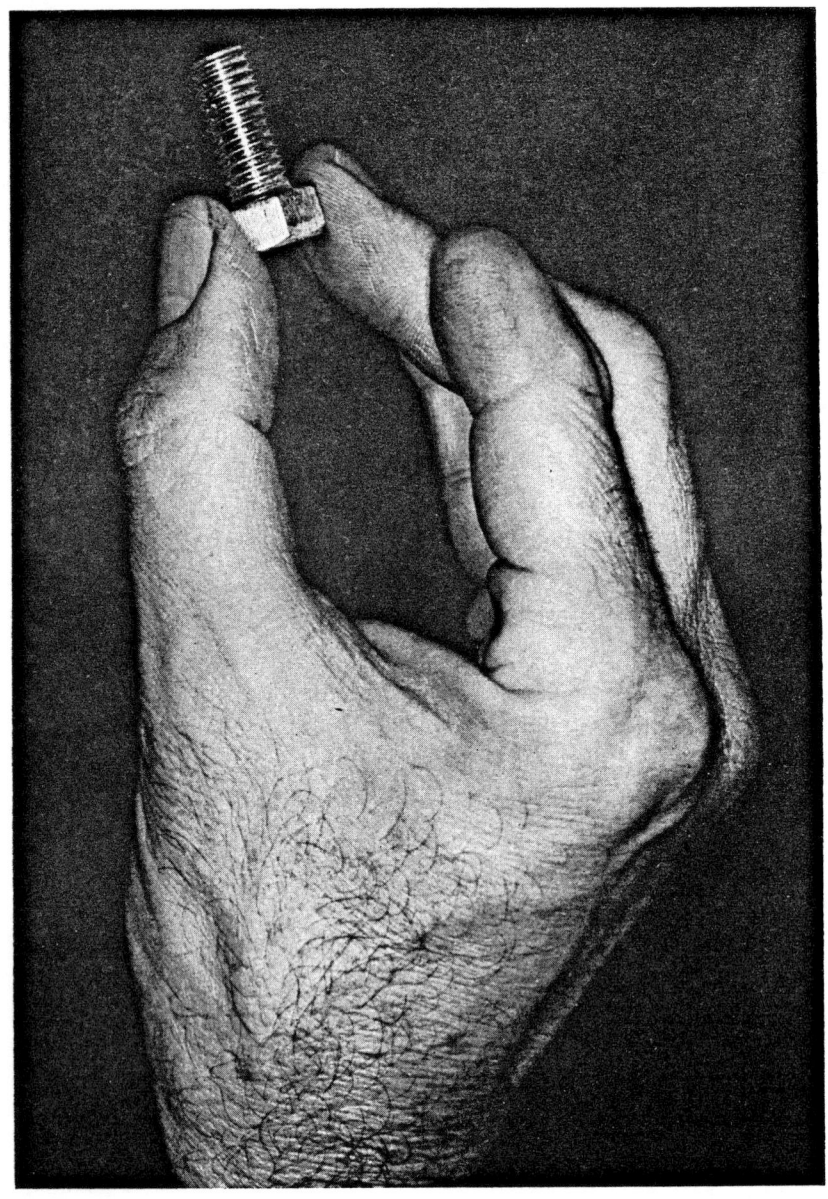

Figure 15. Amputation of the distal phalanx leaves a stump which is not useful for the precise functions of the hand.

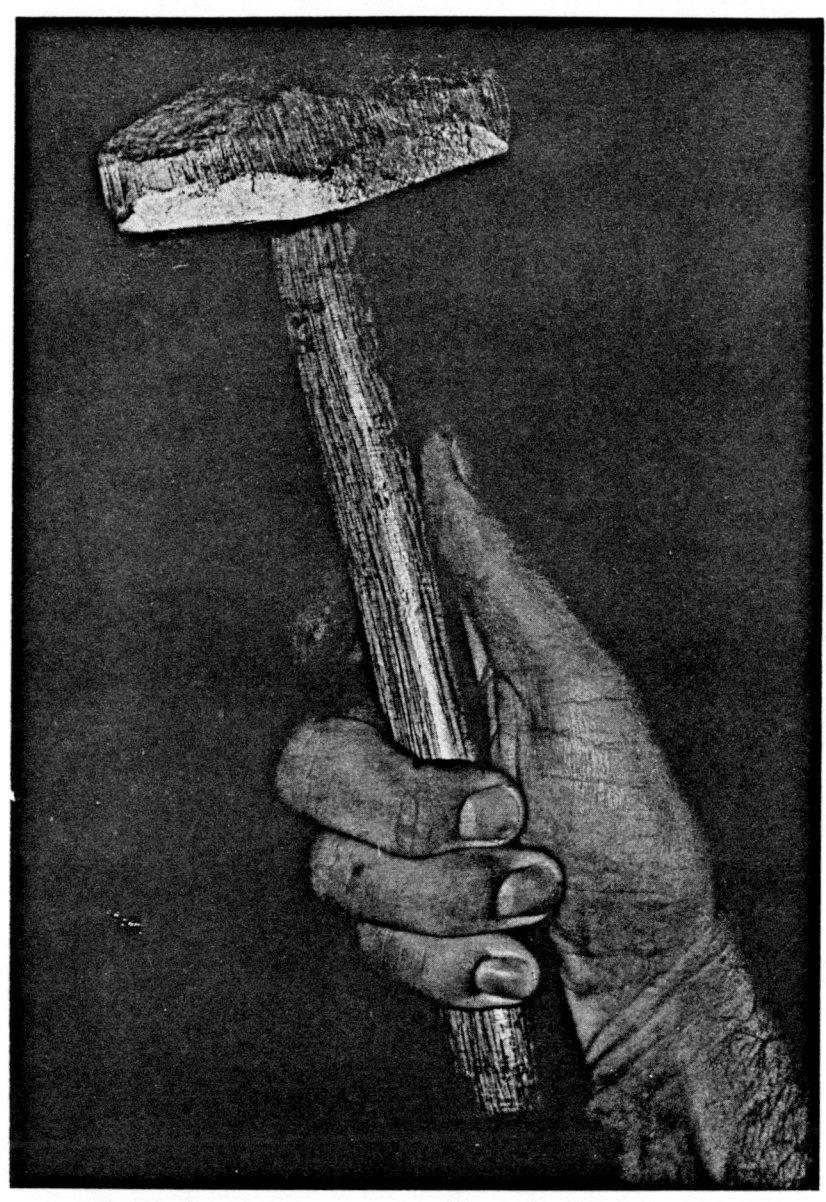

Figure 16. When the distal phalanx has been amputated, the stump is still useful in gross grasp.

days following trauma. Black blood may accumulate beneath the cornified layer of skin. Two or three weeks later this black layer may be peeled off, and a healthy, pink finger may be found beneath the cornified layer.

A completely amputated fingertip is a different problem. Replacement of the entire fingertip succeeds only in small children. It is usually advisable to convert the amputated tip into a free skin graft, removing all the fat and subcutaneous tissue prior to replacement.

SKIN AVULSIONS

Granulating wounds of the hand are invariably accompanied by edema of the part. Edema predisposes the patient to permanent joint contracture (see section on Colles' fracture for a more detailed discussion of edema). Avulsion injuries of the skin of the hand, therefore, must not be permitted to granulate in. Small defects may be closed by undermining the wound margins and suturing. Larger defects must be closed by utilizing pedicle flaps or free skin grafts.

When skin of the hand is incompletely avulsed and remains attached, the severity of the problem may be less obvious. This is especially true if the avulsion occurs at a proximal point and the flap is attached at the distal end of the defect. An avulsion flap which is dependent upon blood supply from its distal end has inadequate circulation, since the arterial flow is primarily from the proximal end. Most avulsion flaps of this type may be expected to undergo at least partial necrosis. In such cases, proper initial treatment is excision of the doomed portion of the flap with primary application of a skin graft or pedicle flap.

HIGH-PRESSURE INJECTION INJURIES

The introduction in industry of high-pressure injection techniques has produced a new type of injury to the hand, an injury which is potentially very serious.

Different substances, such as paint, grease, diesel oil, and molten plastics, may be accidentally injected into the hand at pressures up to several thousand pounds per square inch. The most common site of injury is the volar aspect of a finger. The wound of entrance is a minute hole, and usually there is very little initial pain. The injury may appear to be trivial. Reactive swelling develops within a few hours, and then the pain becomes severe. Often the material will have filled the finger and extended down into the palm. As the foreign material dissects beneath the skin and subcutaneous tissue, the circulation of the skin is compromised. Reactive swelling further compromises and impairs the skin circulation. Only thorough surgical evacuation of the foreign material will prevent skin necrosis. A small incision and drainage at the site of injection is useless. A formal surgical exploration of the finger is essential on the day of the injury. An inadequate or delayed surgical procedure usually results in amputation of the finger.

An injection injury to a finger is illustrated in Figure 17. The patient was initially treated by daily cortisone injections. Not until five days after the injury was the injected paint surgically evacuated, and extensive skin necrosis and secondary infection occurred which made disarticulation at the MP joint necessary. Prompt surgical evacuation of the foreign material on the day of injury probably would have saved this digit.

BURNS

Burns are classified as first degree (erythema), superficial second degree (erythema and blistering), deep second degree (destruction of superficial and intermediate layers of the skin), and third degree (destruction of all layers of the skin and frequent involvement of deeper structures).

First degree thermal burns require no treatment other than the use of an analgesic ointment. Superficial second degree burns respond well to use of an analgesic ointment and a bulky, mildly compressive dressing, combined with continuous elevation of the hand for several days. Although dramatic in appearance (Fig. 18), a large superficial second degree burn is actually more painful than serious. Complete recovery of function can usually be expected within a few weeks (Figs. 19 and 20). In contrast, even a small third degree burn (Fig. 21) can destroy

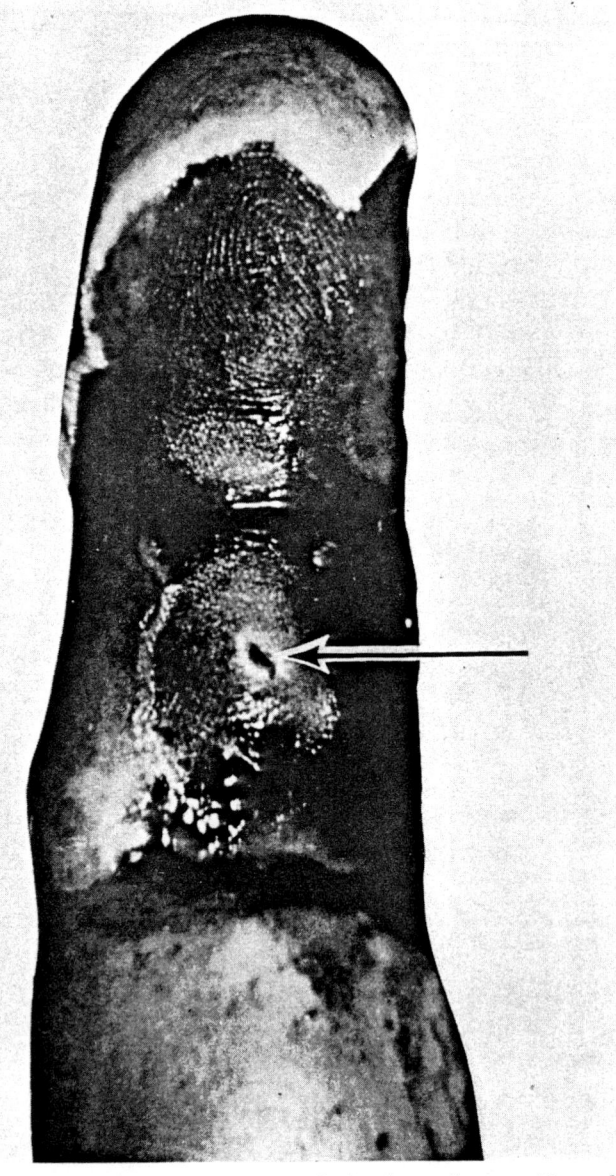

Figure 17. High-pressure injection injury of the long finger. Note the tiny wound of entrance (arrow). There is extensive necrosis of the skin and subcutaneous tissue, a result of incompetent treatment.

not only the skin but also the underlying tendon and bone, which may lead to substantial permanent disability. Deep second degree and third degree burns require expert management.

As a general rule, all third degree burns require skin grafting for three reasons: (a) ungrafted full-thickness skin defects fill in slowly from the edges and large defects require many months to heal; (b) the scar tissue which fills in the ungrafted defect tends to break down with even normal use and contracts as it matures, producing troublesome deformities; and (c) the initial edema produced by the injury persists as long as the wound remains open, and persistent edema predisposes to permanent joint contracture. (See the section on Colles' fracture for discussion of edema.) Therefore, early skin grafting of third degree burns will minimize residual permanent disability.

Chemical burns are first treated by copious irrigation with water. The use of an appropriate neutralizing agent is indicated in some instances. Thereafter, most chemical burns are treated like thermal burns.

Electrical burns may be quite deceptive. Since electrical current can thrombose vessels at a considerable distance from the point of contact, a small surface burn may be accompanied by much more extensive damage deep in the hand, and large areas of gangrene may result. All electrical burns should be promptly referred to a surgeon qualified to render definitive care.

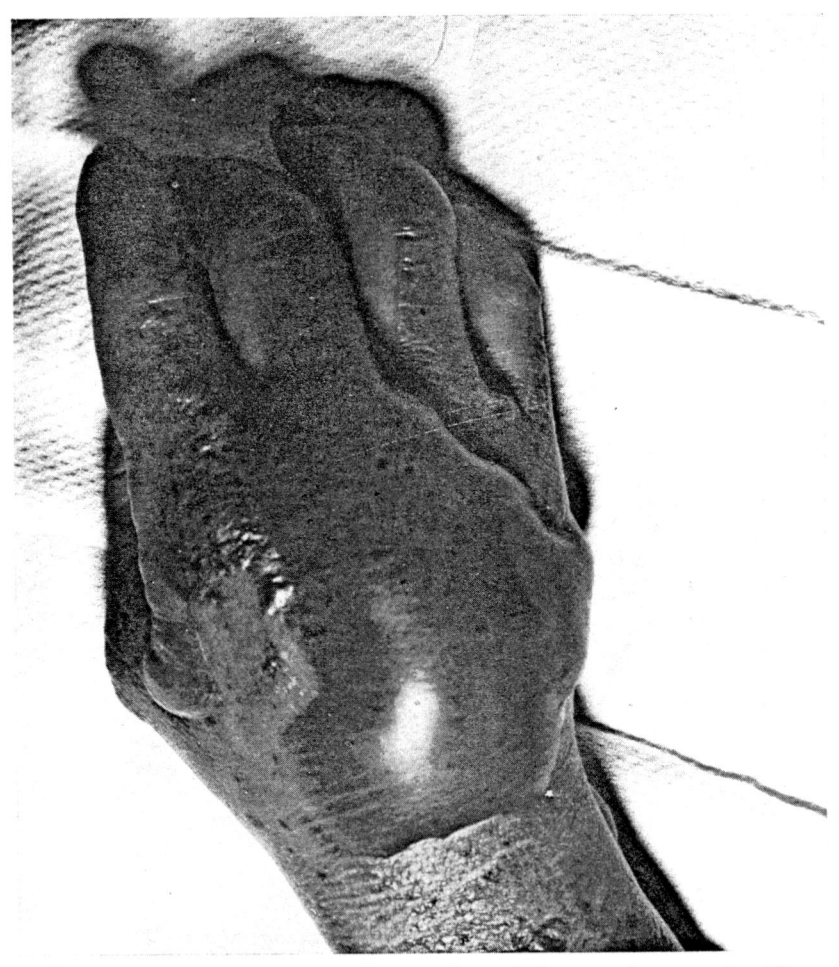

Figure 18. Superficial second-degree thermal burn with extensive blister formation.

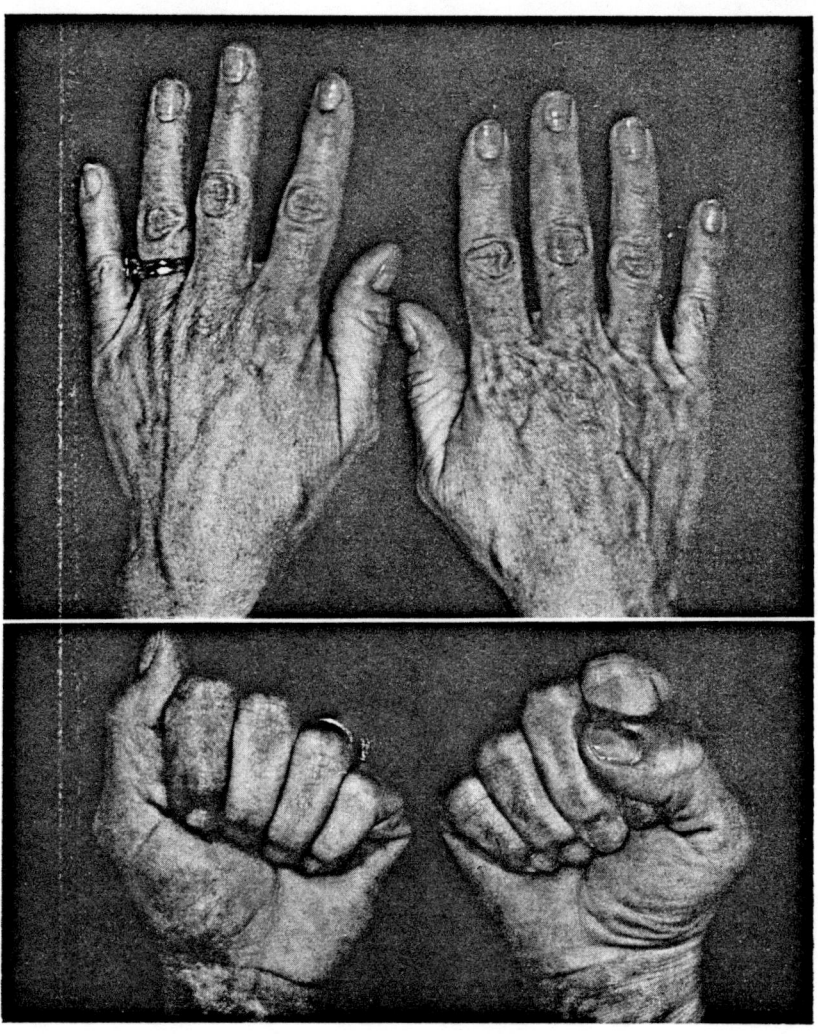

Figures 19 and 20. The patient shown in Figure 18 a few weeks later.

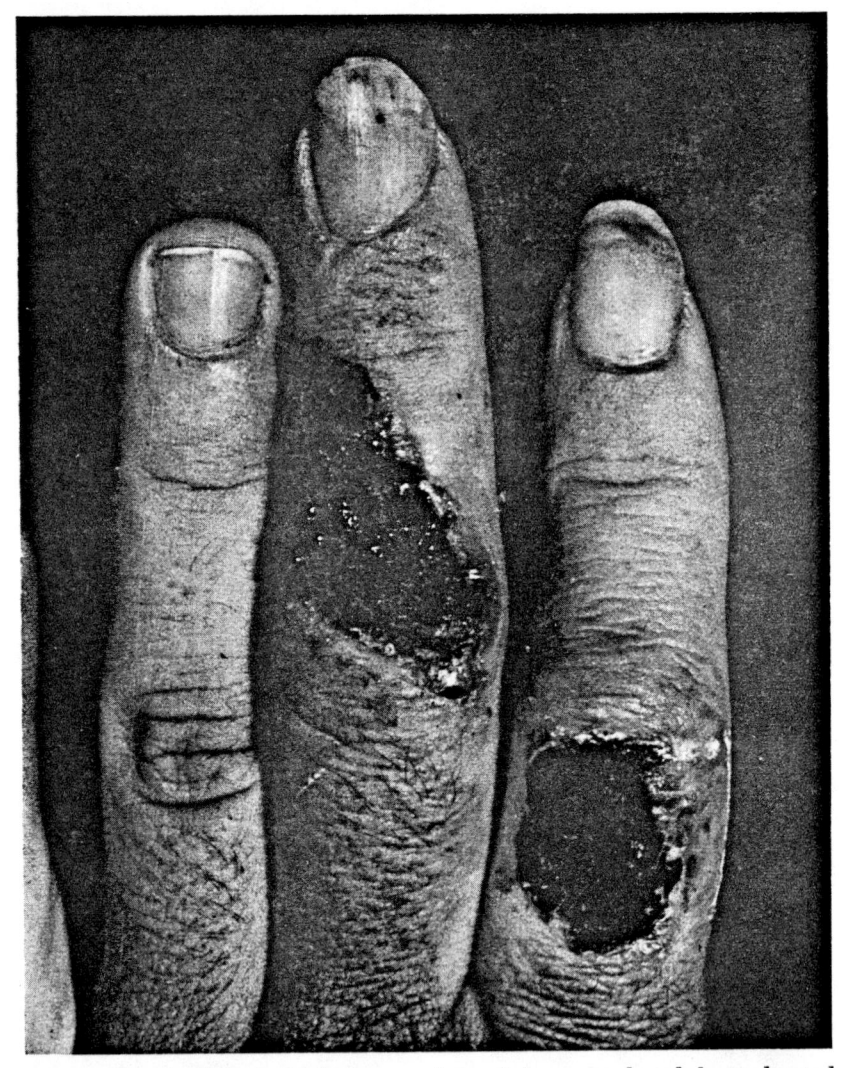

Figure 21. Localized third degree thermal burn produced by a heated press. The underlying extensor tendon and bone were severely damaged.

FRACTURES AND DISLOCATIONS

Colles' Fracture

All too often a well-reduced Colles' fracture which has been immobilized in a cast results in poor hand function, even though appearance of the wrist is good both by clinical and x-ray examination. Conversely, a "skid row" derelict whose Colles' fracture goes untreated may have a permanently deformed wrist, but he regains surprisingly good function of the hand. This apparent paradox is due to violation of one or more of the principles of fracture management.

Four essentials must always be borne in mind by the practitioner:

1. Every effort must be made to prevent edema of the thumb and fingers, and prompt, energetic treatment must be instituted if it occurs. Every fracture, of course, is followed by a period of reactive edema, and this can change a well-fitting cast into one which compromises the circulation of the extremity. A tight wrist cast impairs venous return from the hand and results in edema of the thumb and the fingers. Watson-Jones spoke of persistent edema as "physiologic glue." He was referring to the fibrosis of ligaments and joint capsules and consequent loss of joint mobility which follows persistent edema, and irreversible damage frequently results from only a day or two of edema. The physician should pad the initial cast loosely and generously to allow for the inevitable swelling which occurs at the fracture site. The patient must be instructed to maintain elevation of the hand higher than the elbow continuously until the phase of reactive edema is past; this period may be several days or longer. Whenever the patient complains of pain not relieved by codeine, the practitioner must immediately assume that reactive edema (producing constriction) is the cause of the pain. The cast must be split through the padding to the skin. Fat, edematous fingers are likewise a clear signal that the circulation to the extremity is embarrassed, even if the patient has no complaint of pain. The cast again must be split through all layers to the skin. At times, after splitting the cast, the fracture fragments may slip,

but this loss of reduction must be put into proper perspective. A moderately deformed wrist is frequently only a minor cosmetic problem, whereas stiff digits are a major disability.

2. Prolonged immobilization of joints in nonfunctional positions should be avoided. Permanent loss of joint motion will often result when a major fracture involves a joint, no matter how well the case is managed. This is particularly true in the elderly patient. It is important, therefore, to hold these joints in a position of maximum function if possible. Dorsiflexion is the position of maximum function of the wrist. When the wrist is flexed, the grip is weak and the hand is in a less useful position. If the wrist becomes stiff in the hyperflexed position, the hand

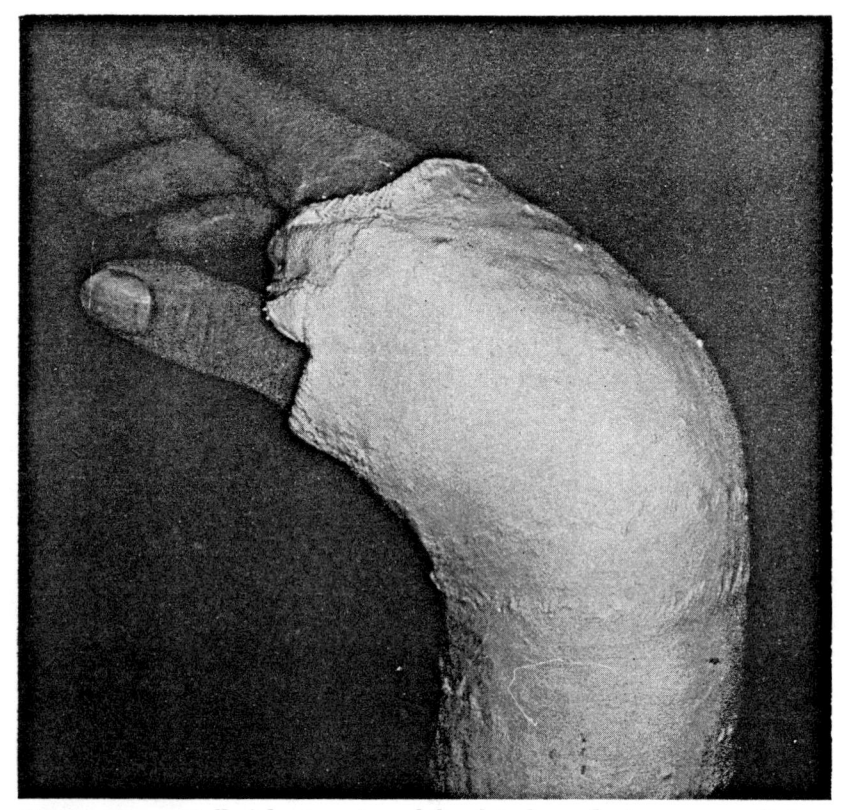

Figure 22. A Colles' fracture immobilized in hyperflexion may result in a crippled hand.

is crippled. Therefore, although most Colles' fractures are most easily reduced and held in hyperflexion, a compromise is necessary. The fracture must *never* be immobilized in the hyperflexed position (Fig. 22). Preservation of function of the thumb and finger joints must take precedence over an anatomic reduction of the fracture.

3. No joint should be immobilized unnecessarily. There is

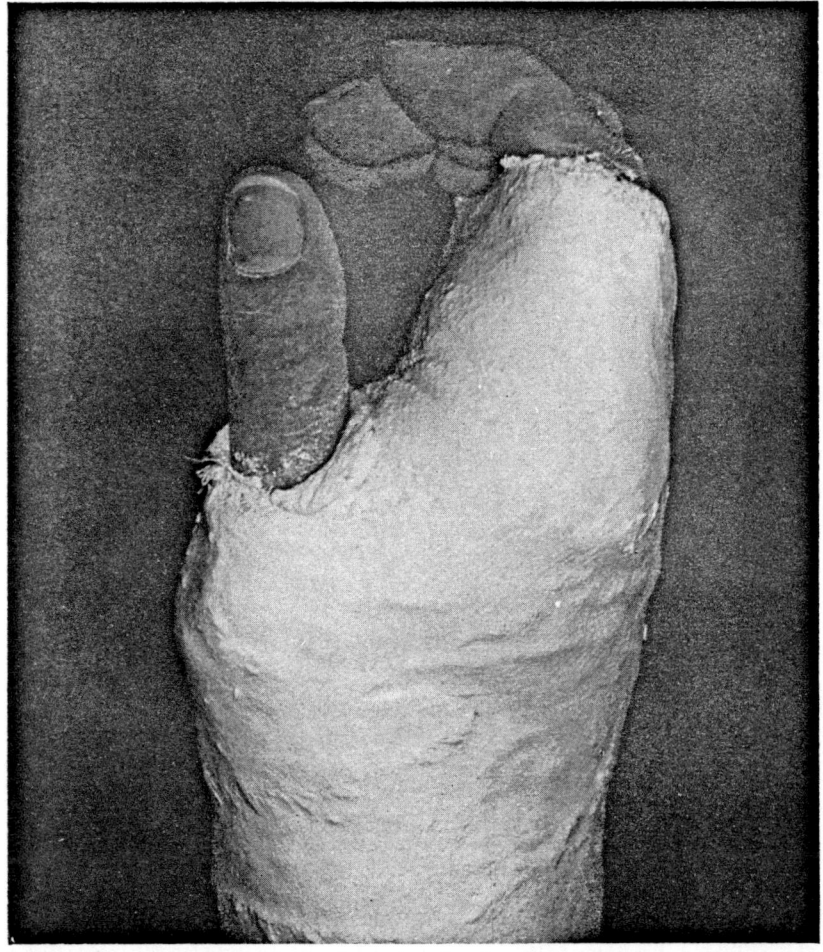

Figure 23. Permanent loss of motion in the thumb and fingers may be caused by a cast which extends out onto the thumb and fingers in adults.

no reason to immobilize the MP or the interphalangeal joints of the thumb or fingers for a Colles' fracture. Casts extending out onto the fingers block motion (Fig. 23). This type of cast may not harm children, but can produce severe permanent stiffness in adult patients. The traditional loop of plaster between the thumb and index finger impairs motion of the MP joints of both the thumb and index finger. The loop is unnecessary. Rotation can be controlled easily by molding the cast carefully in the palm, keeping the thumb abducted away from the index finger.

4. All joints which do not have to be immobilized must be kept actively moving. Active exercises at frequent intervals throughout the day maintain joint mobility and help prevent edema. Each joint must be moved through its full range of motion. The exercises should be active, not passive.

Navicular Fracture

The anatomic snuffbox (Fig. 25) is the triangular hollow at the base of the thumb. It is bounded by the extensor pollicis longus on the dorsal side and by the abductor pollicis longus on the volar side. The radial styloid forms the base of the triangle. The snuffbox overlies the navicular bone. The typical history of a navicular fracture is as follows: The patient, usually a male between the ages of 14 and 40, fell onto his outstretched hand. There was little or no swelling, no deformity, and only a moderate amount of pain. The x-ray examination was negative, and he was told that he had "just a sprain." The pain largely subsided and only months later did he realize that moderate pain about the wrist was still noticeable with moderately active use. Repeated x-rays now may reveal a nonunion of the navicular (Fig. 26). In some cases cystic degeneration is present adjacent to the fracture line (Fig. 27), and in other cases aseptic necrosis may have developed. The treatment of an established nonunion is a bone graft operation. This entails not only hospitalization, but subsequent months of immobilization in plaster. Usually, osteoarthritis of the wrist will gradually develop if surgery is not performed, and arthrodesis of the wrist in order to relieve disabling pain will become necessary in some cases.

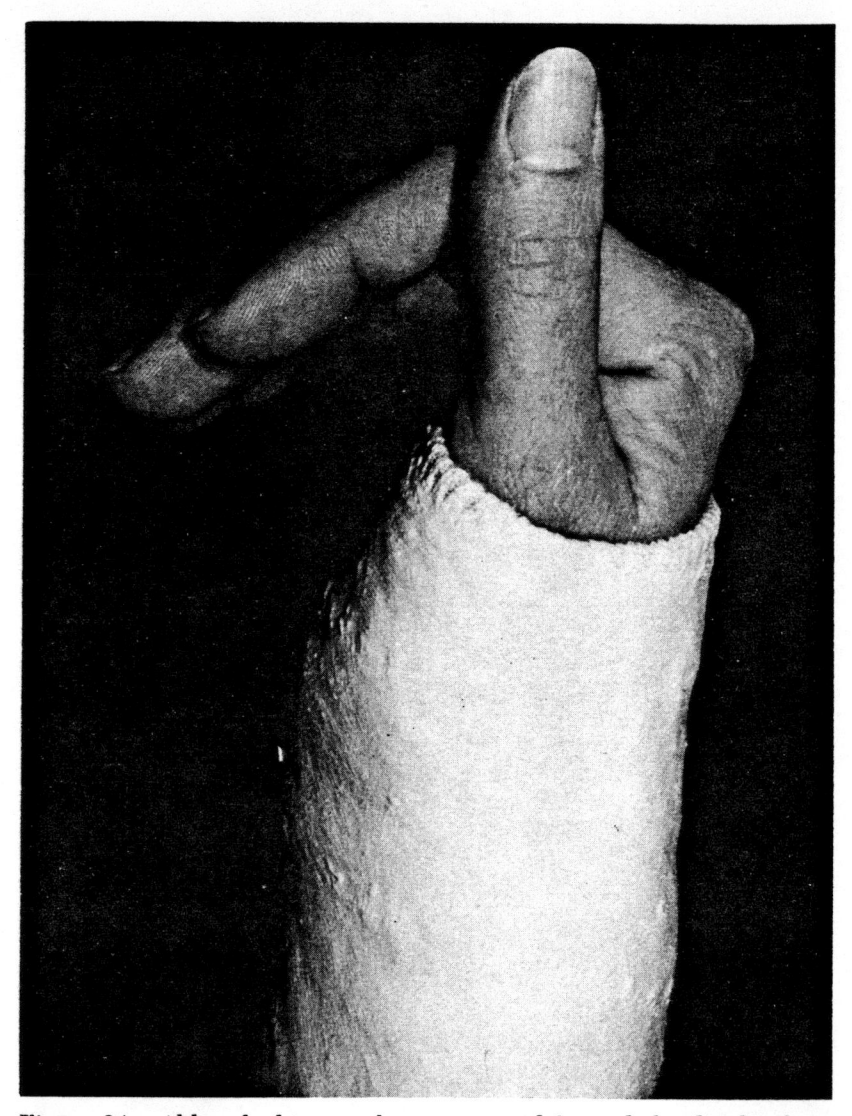

Figure 24. Although this cast does not extend beyond the distal palmar flexion crease, the wrist is held securely. The MP joints can be flexed fully.

The key factor in avoiding nonunion of the navicular bone and its disabling sequelae is early diagnosis. The clinical examination is at least as important as the x-ray examination when the physician evaluates a young adult who has injured his wrist. If there is tenderness in the anatomic snuffbox the physician must

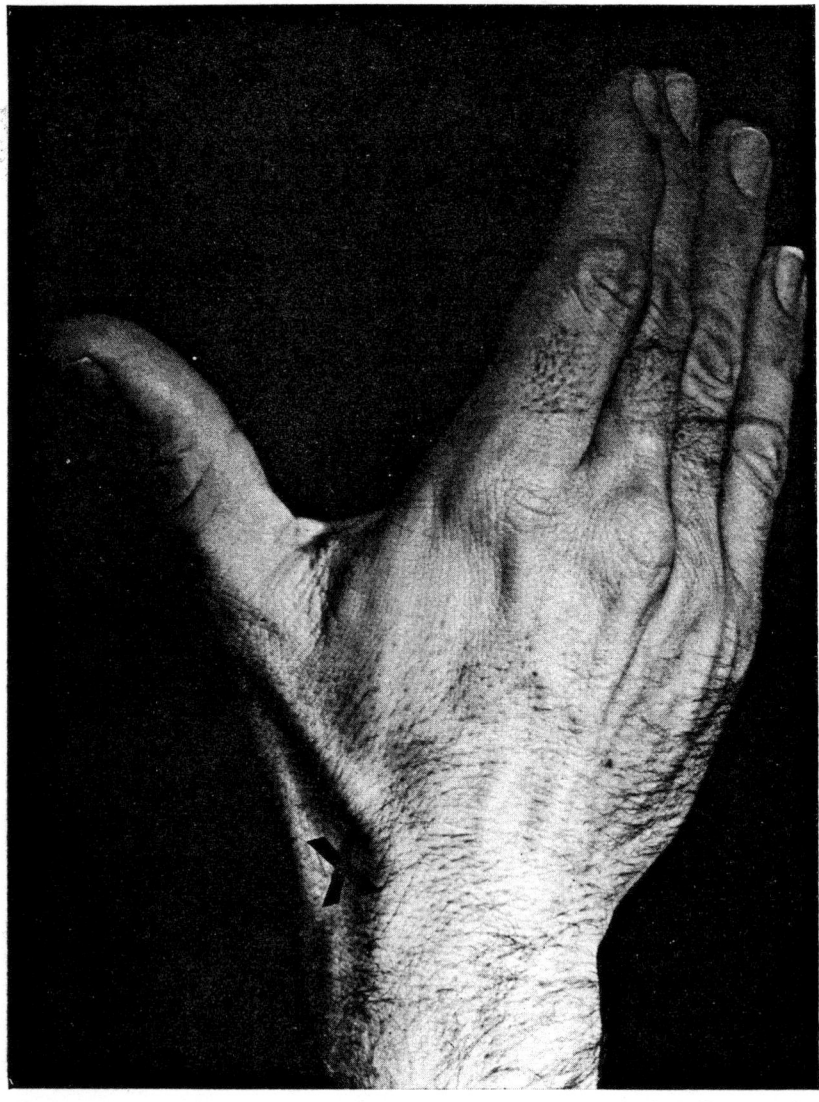

Figure 25. X marks the snuffbox, a triangle bounded dorsally by the extensor pollicis longus, and volarly by the abductor pollicis longus. The radial styloid process is the base of the triangle. Tenderness in the snuffbox suggests a fracture of the navicular.

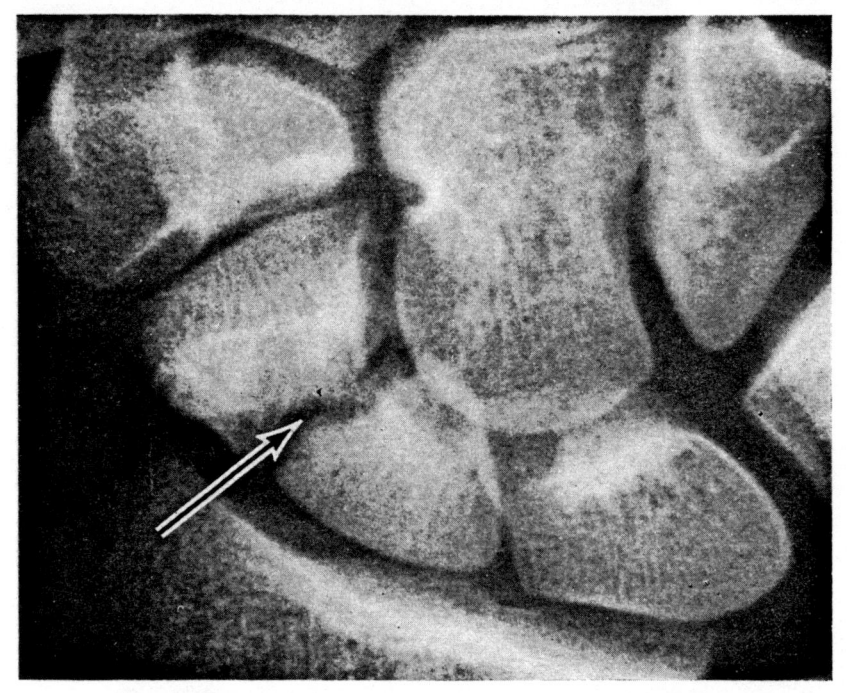

Figure 26. A recent nonunion (arrow) of the navicular.

assume that the navicular has been fractured, despite negative x-ray examination. Apply a cast which includes the proximal phalanx of the thumb, but leaves the fingers free (Fig. 28). Two weeks later remove the cast and repeat the x-rays, obtaining AP, oblique, and navicular views. Multiple views are necessary because a navicular fracture may be visible on only a single view. Since the fracture may be only a fine line (Fig. 29), the x-rays must be of good quality. If the second x-ray examination is also negative, then the injury was indeed "just a sprain" but has nevertheless been well treated.

Unfortunately, it should be recognized that even prompt and proper treatment of a navicular fracture will not always prevent nonunion.

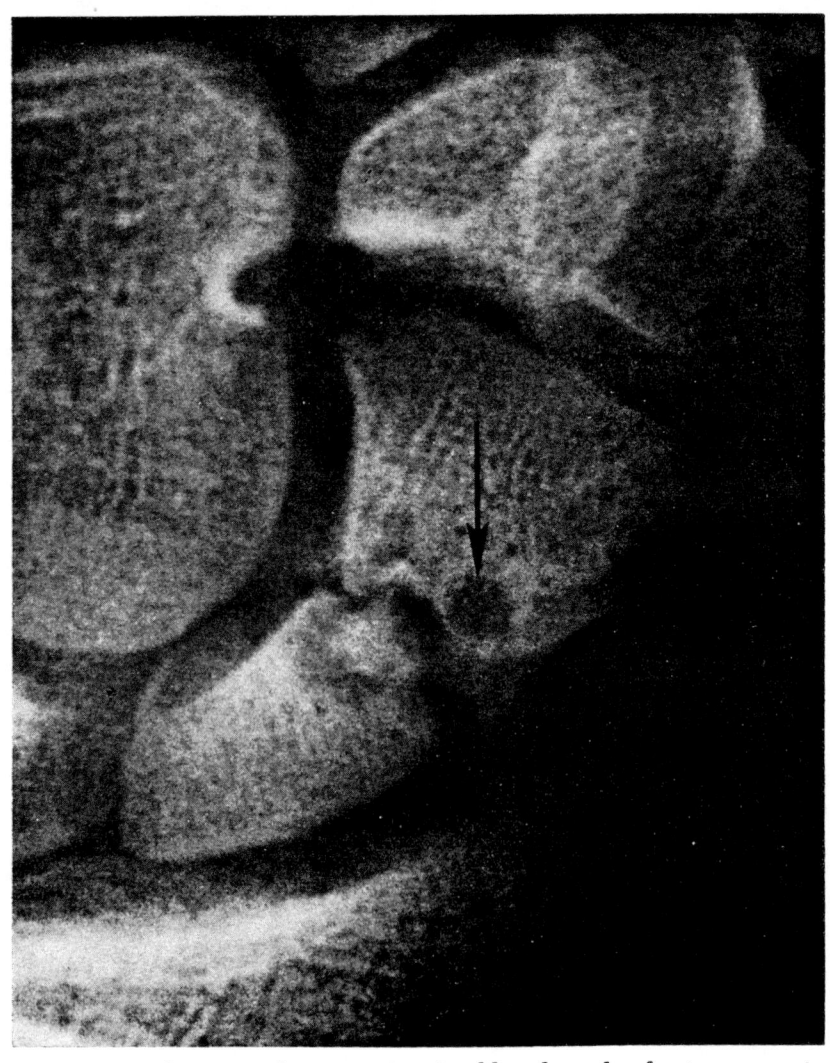

Figure 27. This navicular nonunion is older than the fracture seen in Figure 26. Note the cystic area (arrow).

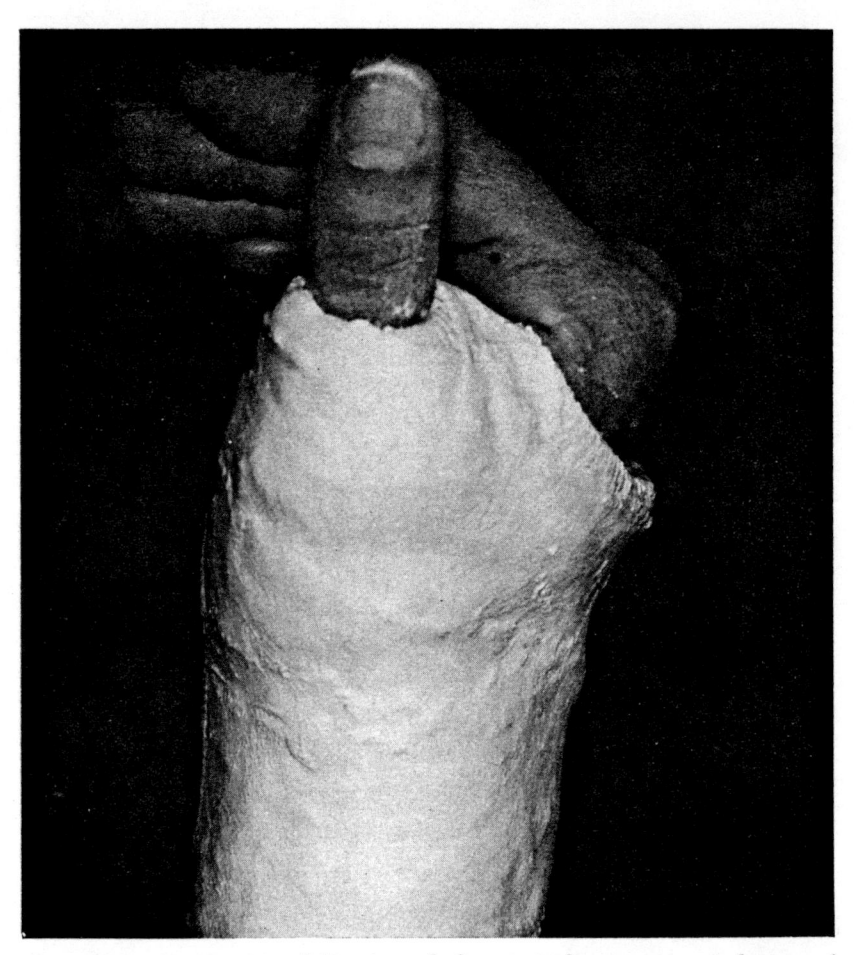

Figure 28. Proper immobilization of the navicular requires inclusion of the proximal phalanx of the thumb in the cast; the fingers are left free.

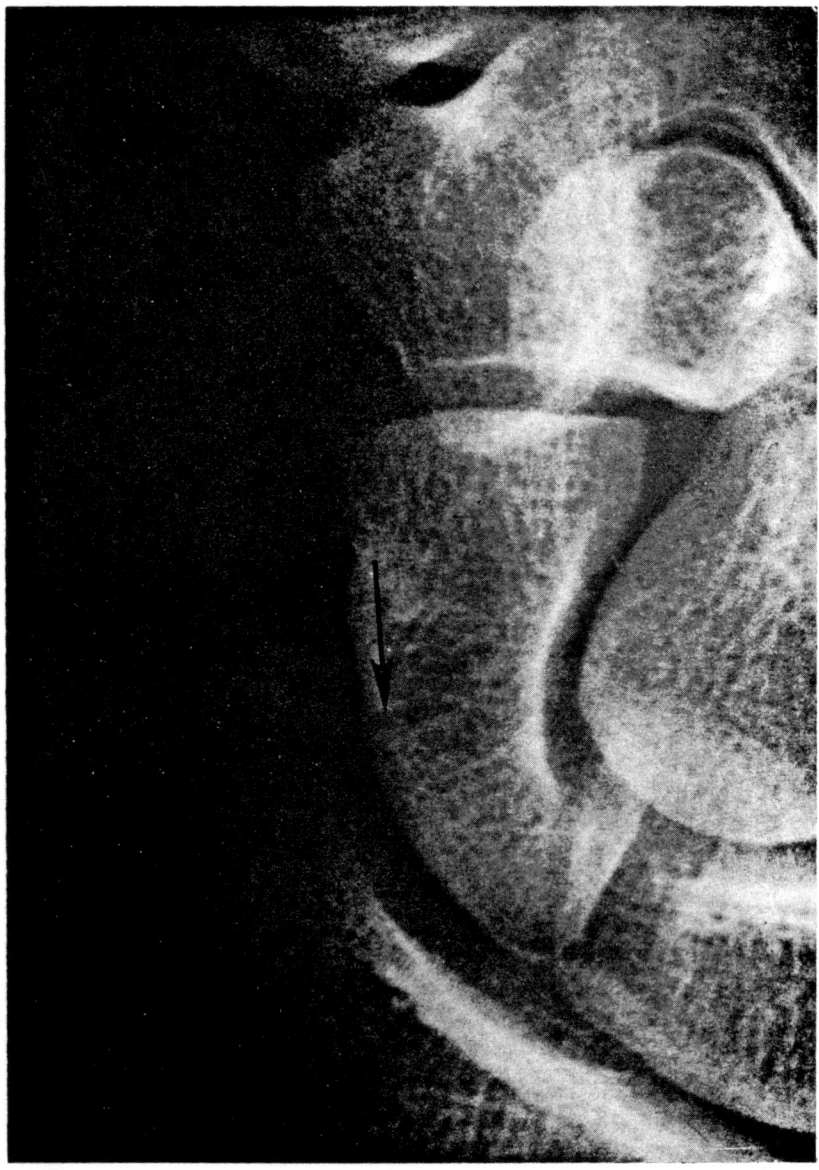

Figure 29. A fine line (arrow) may be the only x-ray evidence of a recent navicular fracture; x-rays must therefore be of good quality.

Navicular Subluxation

Complete dislocation of one or more of the carpal bones produces gross alteration of the bony relationships which are readily apparent on x-ray examination. Although the precise nature of the abnormality may puzzle someone who is not familiar with the radiographic anatomy of the hand, it should be obvious even to inexperienced observers that something is amiss. These dislocations are caused by major trauma and are frequently accompanied by fracture of the navicular bone. Such injuries are serious, difficult to treat, and frequently result in substantial permanent disability.

In contrast to complete dislocations, rotary subluxation of the navicular involves only minor degrees of displacement and is therefore easily overlooked. Subluxation of the navicular is a serious injury. If the lesion is not treated, the patient will be left with a wrist which is painful and weak, with limited motion. Navicular subluxation commonly follows a fall onto the outstretched hand. Pain may be modest and swelling minimal. When examining x-rays, the physician often will not notice the minor displacement of the navicular.

The most important x-ray signs are (a) widening of the space between the navicular and the lunate and (b) loss of height (apparent shortening) of the navicular as seen in the AP view. Widening of the space between the navicular and the lunate (Figs. 30 and 31) is most reliably demonstrated on an AP x-ray view of the *supinated* wrist (palm up). Identical comparison views of the opposite wrist should be taken in suspicious cases.

Although the displacement of the navicular is minimal, open reduction is often necessary. Internal fixation by Kirschner wire is usually necessary to maintain reduction. *Significant permanent disability* follows anything less than an anatomical reduction.

Bennett's Fracture-subluxation

Injuries to the base of the thumb can produce a Bennett's fracture-subluxation of the first carpometacarpal joint. The fracture fragment is usually small, and the amount of displacement may be only a few millimeters (Fig. 33). This lesion may

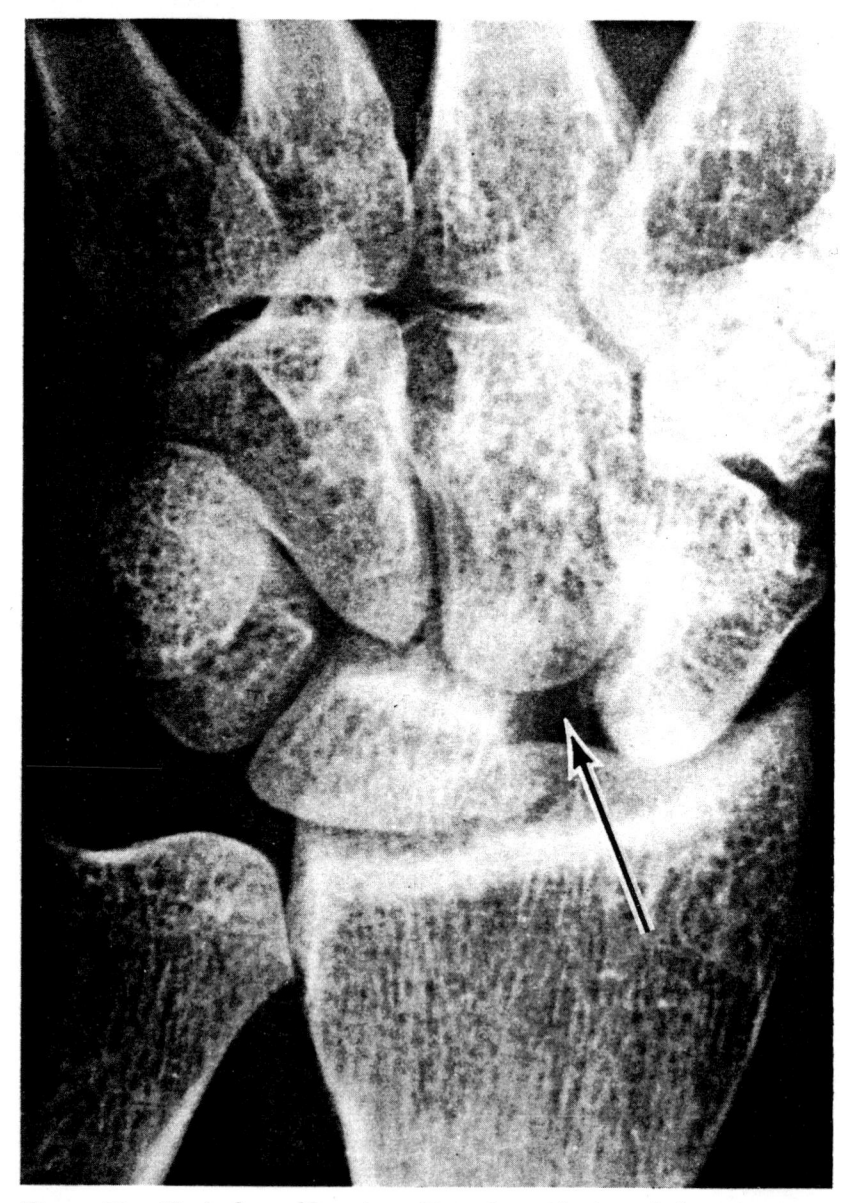

Figure 30. Navicular subluxation. Note the widening of the joint space between navicular and lunate (arrow) and the apparent shortening of the navicular. Compare with the normal hand shown in Figure 31.

Acute Injuries 45

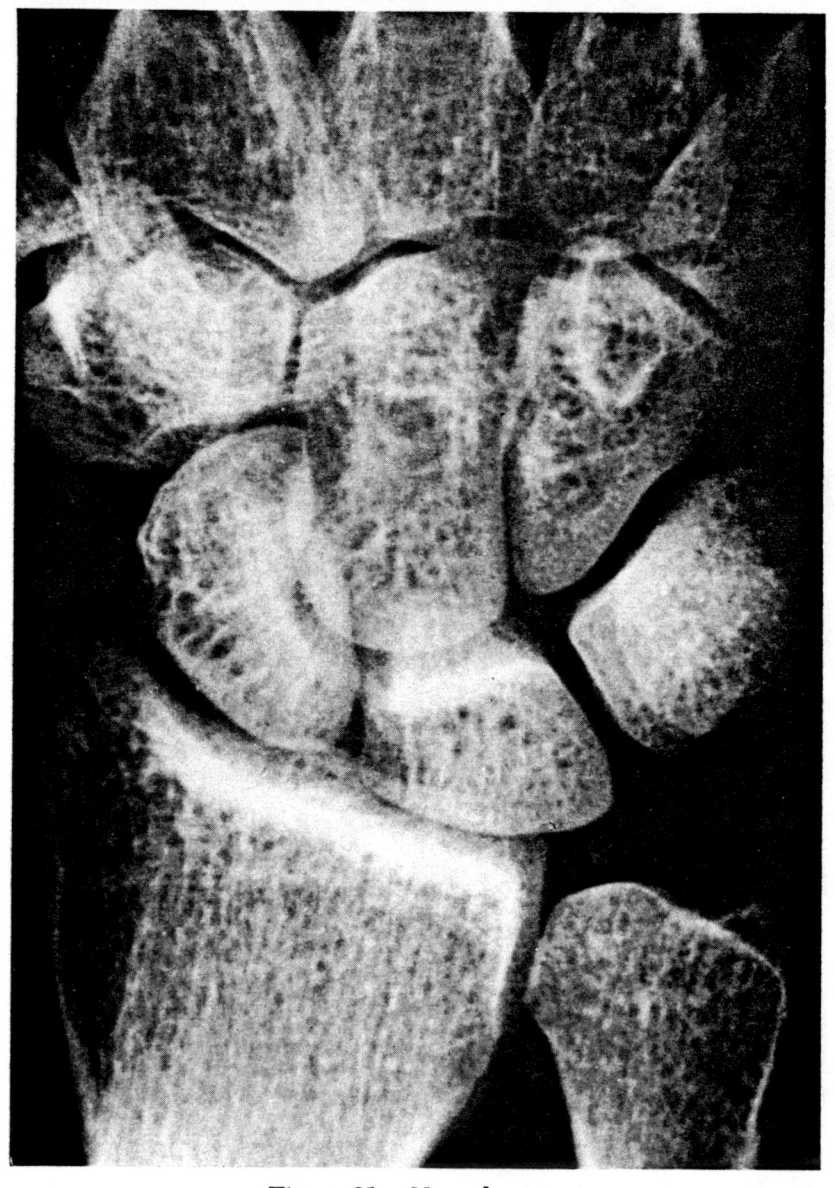

Figure 31. Normal wrist.

46 · *The Hand Book*

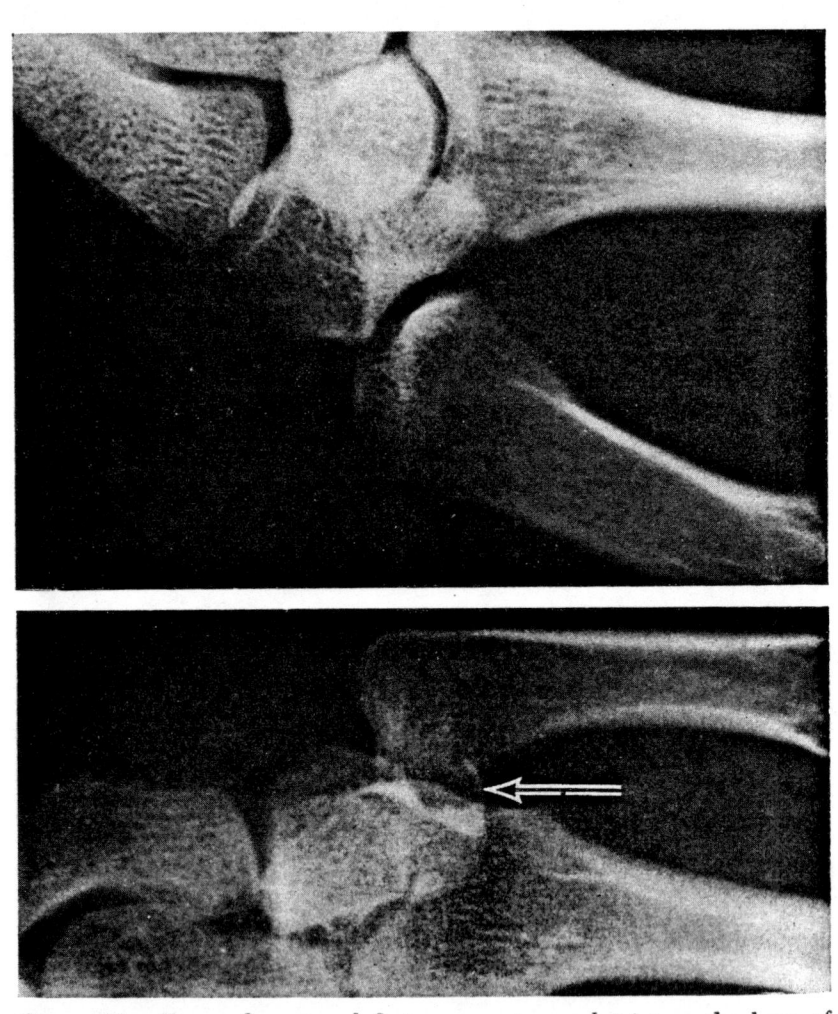

Figure 32. X-ray of a normal first carpometacarpal joint at the base of the thumb.

Figure 33. Bennett's fracture-subluxation. The first metacarpal is displaced away from the second metacarpal. A small fragment of bone (arrow) is fractured off the first metacarpal.

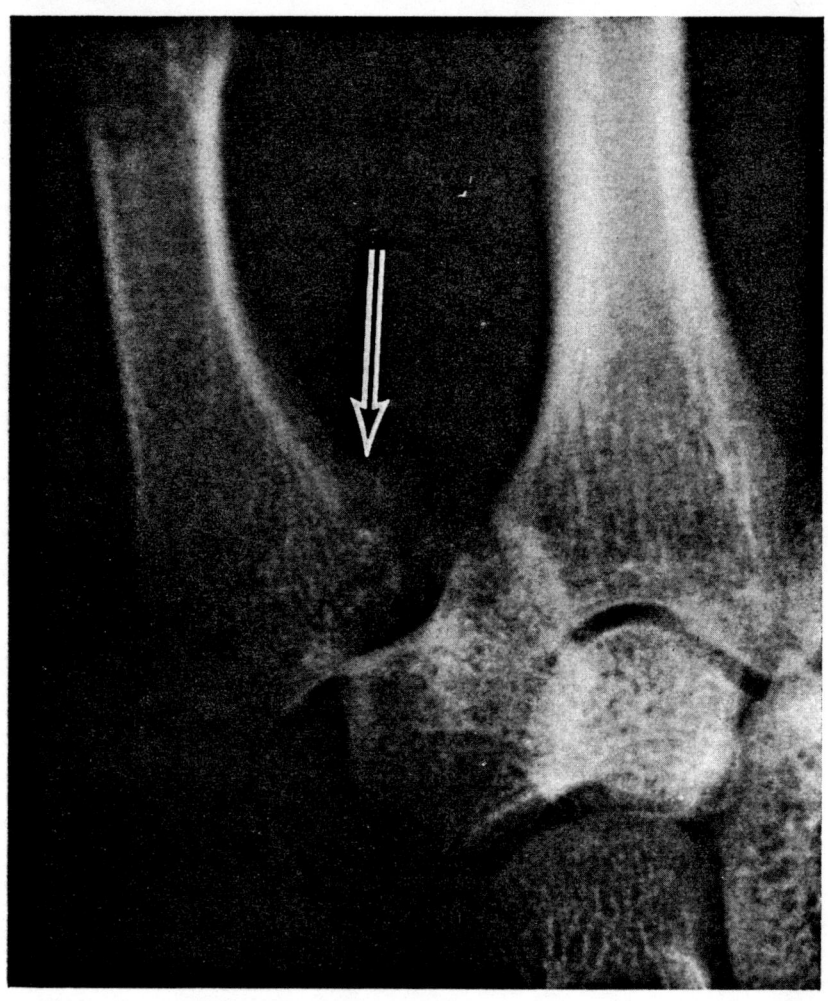

Figure 34. A mistreated case of Bennett's fracture-subluxation seven months after injury. Note the arthritic spur (arrow).

easily be dismissed as "just a chip fracture." However, anything less than an anatomic reduction will result in osteoarthritis, which can disable the thumb. Anatomic reduction almost always requires Kirschner wire fixation and usually an open reduction.

The patient whose x-ray is illustrated in Figure 34 was initially treated by simple cast immobilization. Obvious osteoarthritis is demonstrated on the x-ray taken only seven months after injury. The patient is now totally disabled for his occupation as a self-employed barber. Fortunately, arthrodesis of the arthritic joint will eliminate pain and restore strength to the thumb.

Metacarpal Fracture

The most common metacarpal fracture occurs at the neck of the fifth metacarpal bone. This injury is sometimes called the "boxer's fracture," since it often is incurred by striking an opponent's head or the wall behind. The diagnosis can be suspected from the history and the location of the pain. Swelling is also commonly present, and the fracture is readily confirmed by x-ray examination (Fig. 35). Overzealous strapping of the fingers in the fist position in order to maintain a reduction can produce a poor functional result because of finger joint contractures. On the other hand, incomplete reduction of this fracture may produce a noticeable deformity, but will result in little or no functional disability.

Fractures of the shafts of the metacarpals are not often significantly displaced and usually require no more than simple immobilization. However, a shaft fracture which is significantly displaced will require reduction (Fig. 36). An unreduced overriding fracture of a metacarpal shaft may result in a bony prominence in the palm, which will seriously limit the ability to grasp. A transverse shaft fracture which is overriding can usually be managed by closed manipulation and a cast, but an oblique or comminuted fracture may require closed pinning or open reduction and pinning.

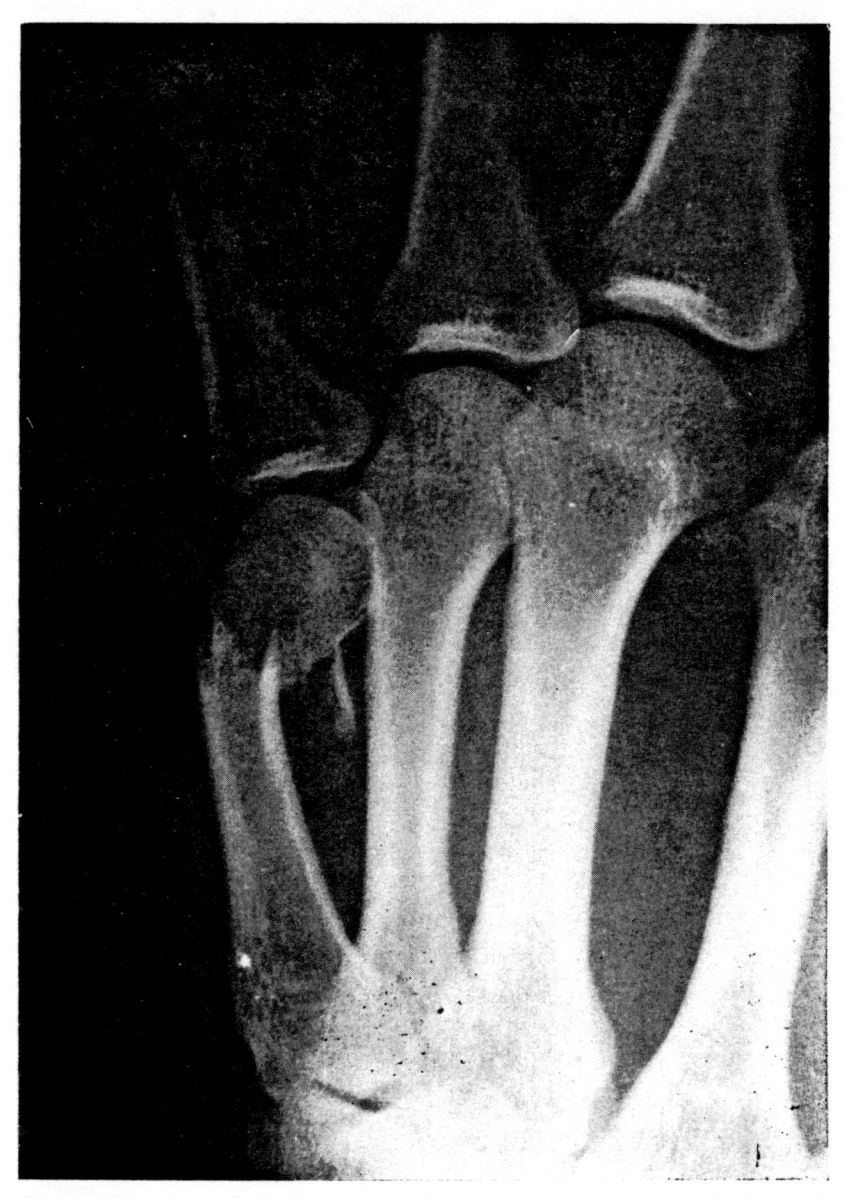

Figure 35. A moderately angulated fracture of the neck of the fifth metacarpal.

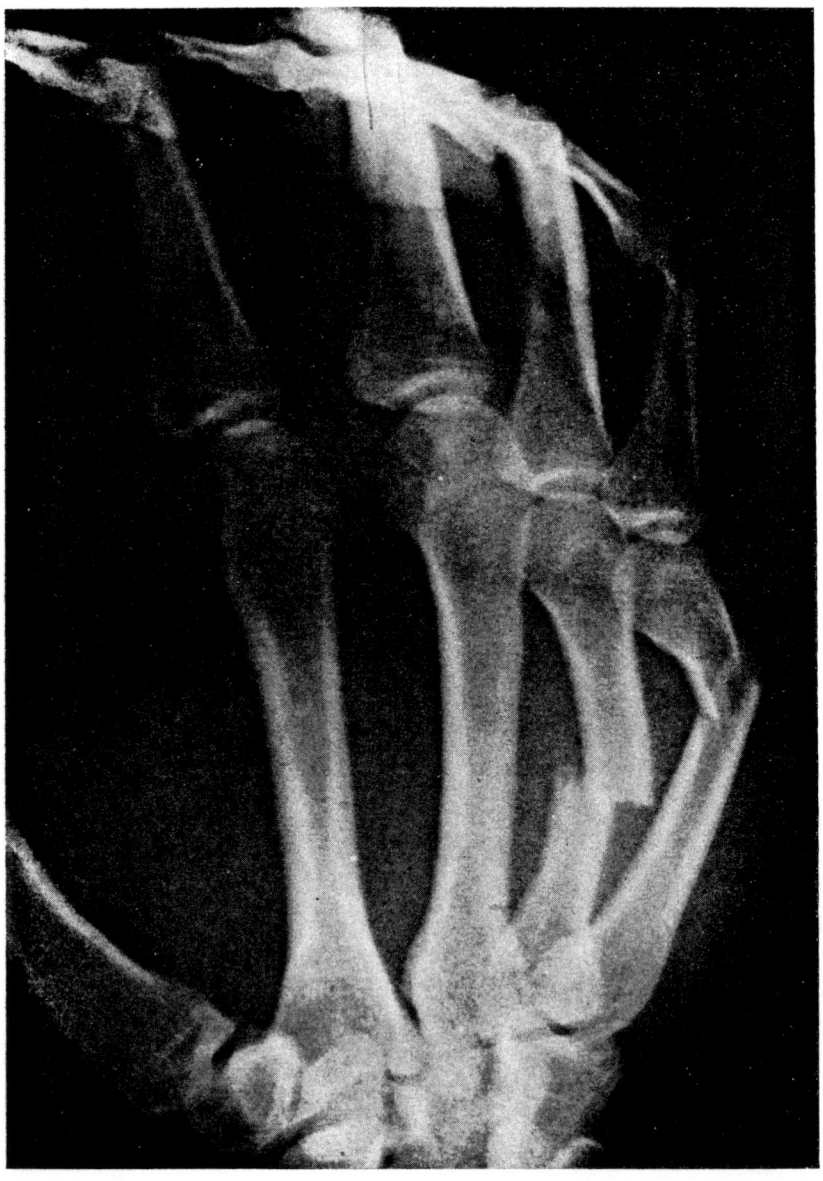

Figure 36. An overriding fracture of the shaft of the fourth metacarpal, with an angulated fracture of the neck of the fifth metacarpal.

Metacarpophalangeal Dislocation

The most frequently dislocated MP joint is the MP joint of the thumb. Many of these dislocations are easily reduced by only token manipulation and can then be immobilized. If the metacarpal head has buttonholed through a rent in the capsule of the joint, the dislocation stubbornly resists reduction. Such a dislocation cannot be reduced except by surgery. The dislocation shown in Figure 37 could not be reduced by manipulation under general anesthesia. Operation revealed a buttonhole tear of the capsule.

Phalangeal Fracture

An injury to or near a finger joint frequently produces fibrosis and consequent stiffness of the joint. Stiffness may even occur in fingers adjacent to the injured digit. Such stiffness develops most frequently in the middle-aged or older patients and when there is marked edema (see section on Colles' fracture). No joint should be immobilized unnecessarily. Contractures are less likely to develop in those joints which are left free so that the patient can move them through a full range of motion at regular intervals throughout the day.

A basic principle of fracture management is that both the joint proximal to the fracture and the joint distal to the fracture must be immobilized. Usually the middle and distal phalanges may be adequately immobilized on a finger splint. Fractures of the proximal phalanx, however, require immobilization of the MP joint. This is most effectively done by applying a cast to the hand and lower forearm and incorporating a metal finger splint. The finger is then taped to the metal splint (Fig. 38).

The collateral ligaments of the MP joints are lax when the joints are extended, and tight when the joints are flexed. If fibrosis of a collateral ligament occurs when an MP joint is immobilized in extension, the patient will be unable to flex the joint. Whenever possible, therefore, the MP joints should be immobilized in the flexed position. The PIP joints, on the other hand, are prone to develop flexion contractures, and the PIP joints should be immobilized in extension if possible.

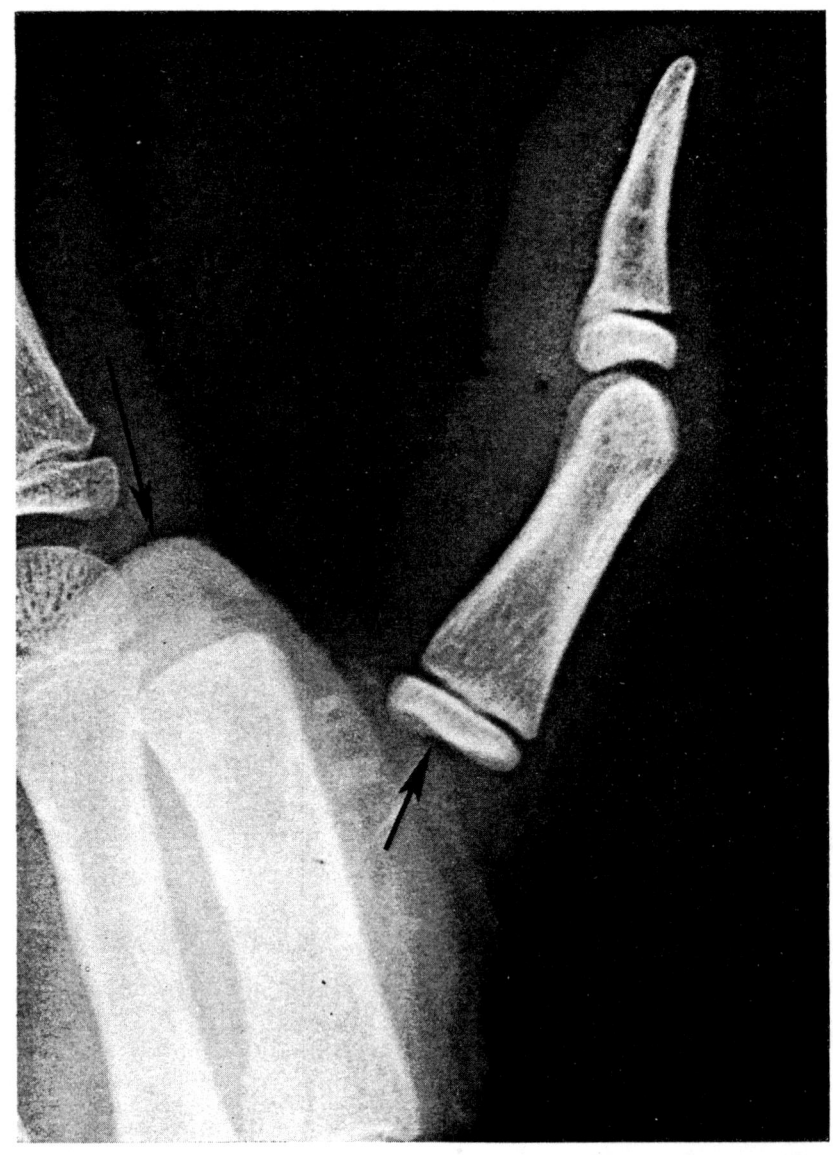

Figure 37. Dislocation of the MP joint of the thumb in a child. The proximal phalanx (short arrow) is displaced from the metacarpal head (long arrow).

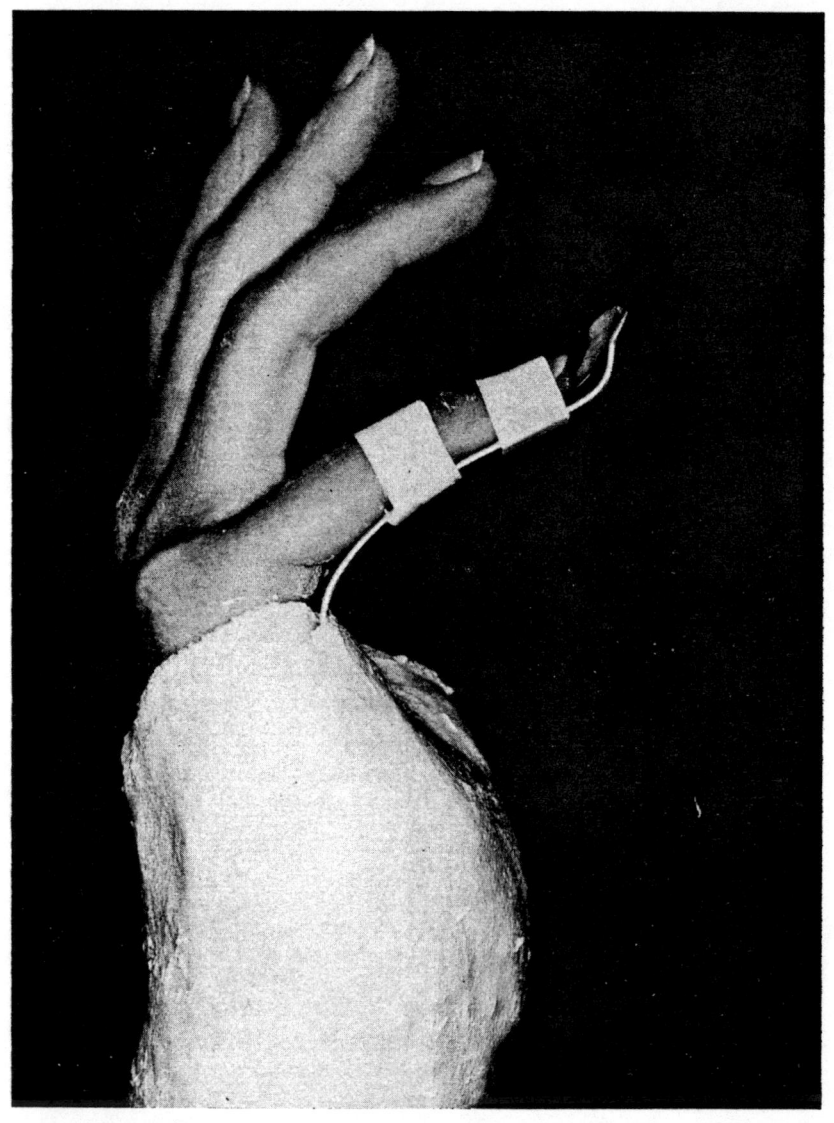

Figure 38. Fracture of the proximal phalanx is immobilized in a forearm cast incorporating a metal splint.

54   *The Hand Book*

When fractures of the shaft of the proximal or middle phalanges occur, the presence or absence of overriding and the amount of angulation deformity can be readily determined from the x-rays. However, rotational deformities are not well demonstrated by x-ray and must be looked for by clinical examination. This examination is based on the fact that although the fingers tend to lie apart from each other when extended (Fig. 39), they crowd together when flexed (Fig. 40). Therefore, after manipula-

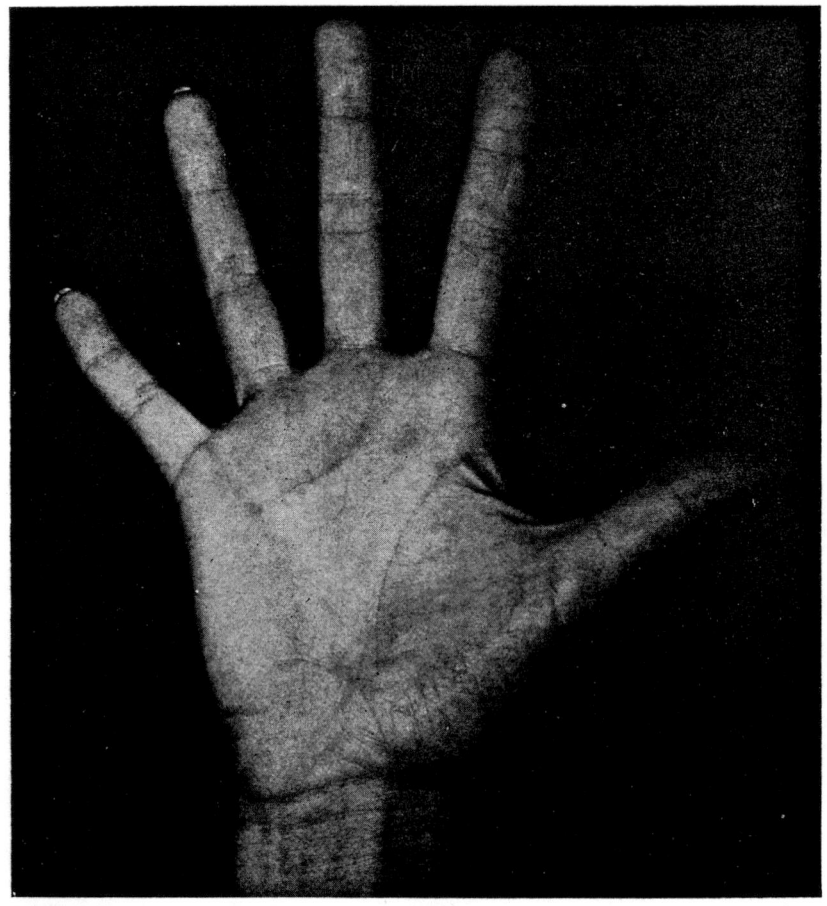

Figure 39. The fingers of the normal hand tend to spread apart when extended.

tion of an obviously displaced fracture or in evaluation of a fracture which by x-ray appears to be undisplaced, have the patient flex his fingers so as to bring the fingertips together on the palm. If there is significant rotational deformity, the involved finger will tend to lie under or on top of its neighbor. Rotational deformities usually can be corrected easily by grasping the tip of the finger and overcoming the deformity prior to taping the finger to a splint.

Fractures of the phalanx which involve a joint require anatomic reduction of all but the very smallest fragments. Open

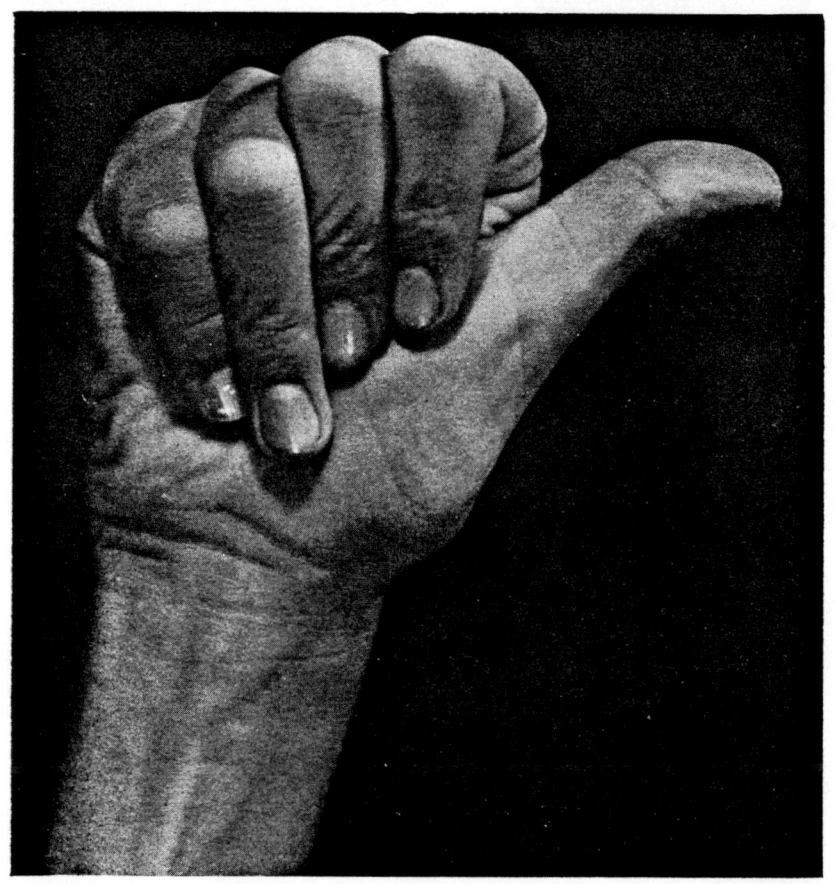

Figure 40. The fingers of the normal hand crowd together when flexed.

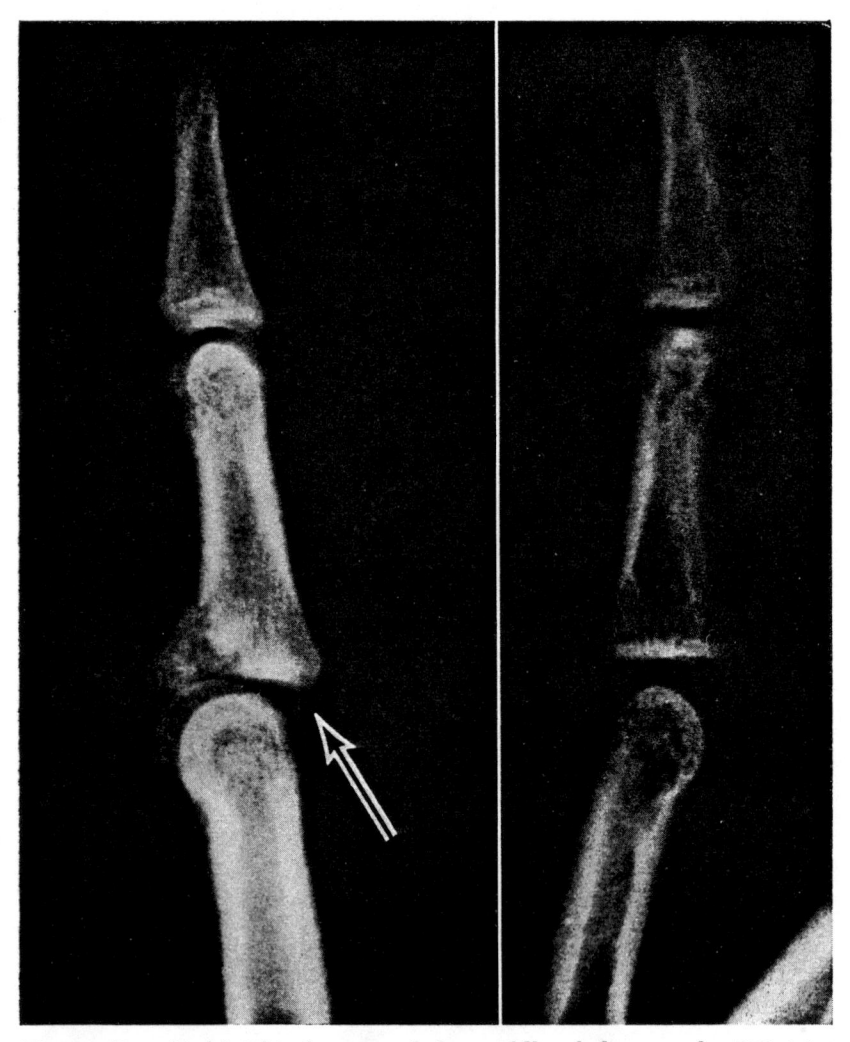

Figure 41. (Left) This fracture of the middle phalanx at the PIP joint produces dorsal subluxation (arrow) of the middle phalanx on the proximal phalanx. Compare with Figure 42.

Figure 42. (Right) Normal finger, lateral x-ray view.

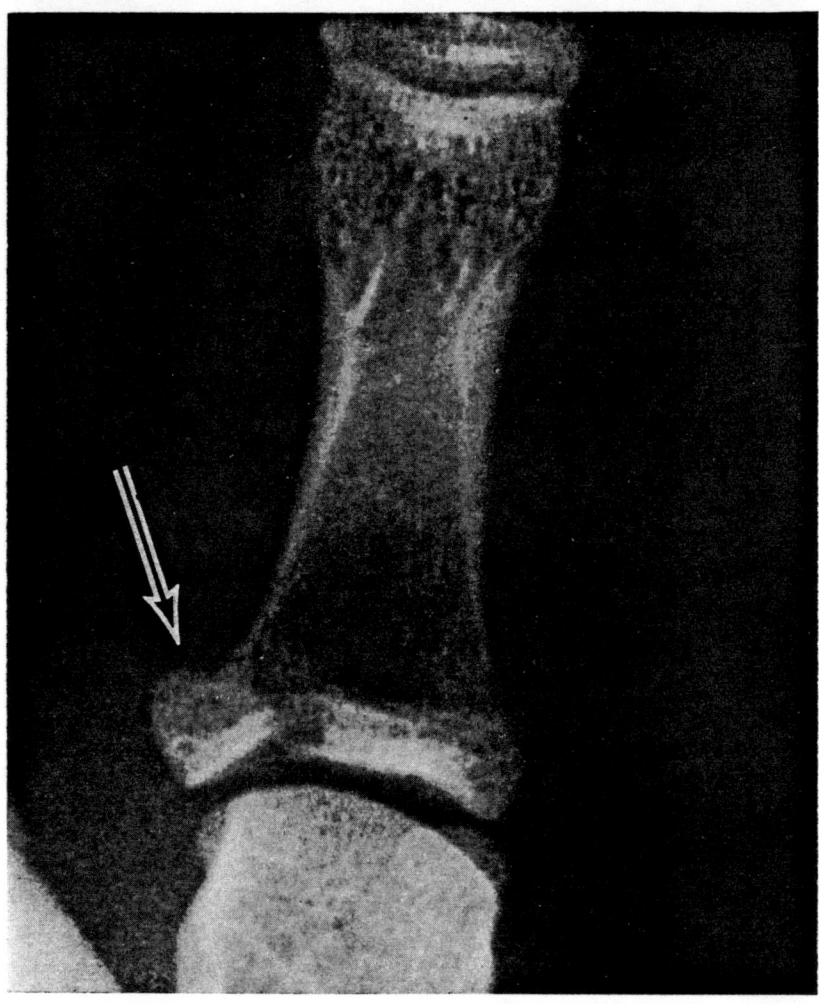

Figure 43. Fracture of the thumb at the MP joint. The collateral ligament attaches to the fragment (arrow). The joint will be unstable if the fragment is not replaced.

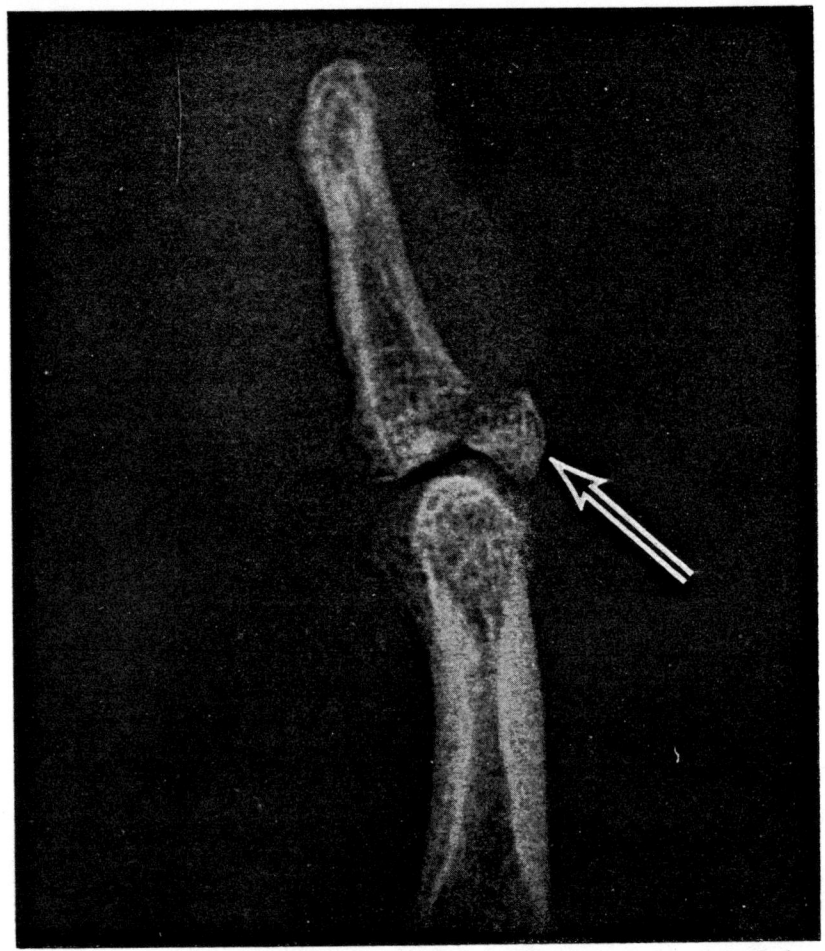

Figure 44. Avulsion fracture of the DIP joint. The fragment which has been pulled loose is the bony insertion of the extensor tendon.

reductions are sometimes necessary. This is especially true if the fracture fragment is large enough to result in subluxation of the joint (Fig. 41). If there is fracture of an important ligamentous attachment, as illustrated in Figure 43, open reduction is likewise necessary. Unfortunately, the joints of the hand do not tolerate injury well; consequently, the adult patient may not regain a normal range of motion.

Closed fractures of the distal phalanx are rarely troublesome and usually require only simple immobilization on a splint, pro-

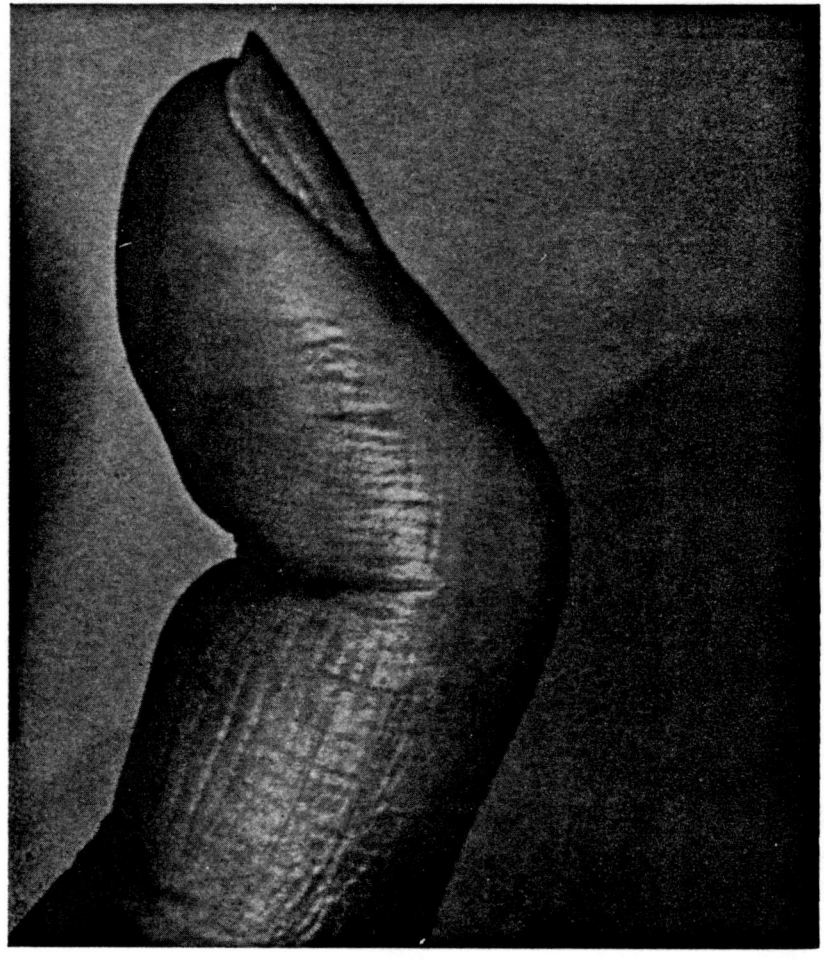

Figure 45. Mallet finger deformity.

vided that the joint is not involved. The most common fracture involving the DIP joint is an avulsion fracture, in which the extensor tendon pulls loose its bony insertion from the distal phalanx (Fig. 44). The patient is unable to extend the DIP joint fully. The resultant flexion deformity is called a mallet finger (Fig. 45). If the fragment involves 50 percent or more of the articular surface, the joint can sublux, and open reduction is mandatory. If a flexion deformity of over 30° is present, this deformity is unacceptable since it interferes with normal function of the hand. Flexion deformities of less than 30° are of no functional significance. Some women will find them unacceptable cosmetically, however. If the deformity is unacceptable functionally or cosmetically, open reduction will be necessary in most cases since accurate closed reduction is rarely possible. Osteoarthritis will develop in some patients whether or not surgery is performed. Subsequently, arthrodesis may become necessary to relieve joint pain.

Rupture of the extensor tendon near its insertion into the distal phalanx will produce a mallet finger deformity similar in external appearance to the avulsion fraction of the DIP joint. Since rupture of the extensor tendon does not involve injury to the joint, osteoarthritis rarely follows this injury. If the flexion deformity is less than 20°, no treatment is necessary. Joint pain will gradually disappear over a period of several months. If the flexion deformity is over 20°, or if the patient finds the deformity unacceptable from a cosmetic standpoint, the finger should be splinted for four weeks. Immobilize the DIP joint in maximum hyperextension and the PIP joint in 45° of flexion. All but the lightest pressure should be avoided on the dorsum of the PIP joint, as this area of the skin of the finger is particularly prone to develop pressure necrosis. Skin necrosis in this location is a far more serious problem than a mallet finger deformity. Even after treatment by immobilization, some residual flexion deformity may result. If the patient finds the residual deformity unacceptable, surgical correction can be carried out.

REFLEX SYMPATHETIC DYSTROPHY

Reflex sympathetic dystrophy usually arises as a result of a minor injury. It occurs most frequently in middle-aged women. Typical is the crushing injury to a fingertip sustained by closing the tip in a car door. Often the bone is not fractured, and initially the injury may be considered trifling, but the patient may note increasing pain and local swelling. The skin of the fingers and hands becomes dystrophic and shiny and the normal skin creases disappear. The painful area may progressively increase in size, and the patient begins to favor the hand. Commonly the hand is darker and cooler than the normal hand, and there may be increased diaphoresis. Occasionally the hand may be warmer and appear red as compared to the uninvolved hand, or there may be no temperature or color difference between the two hands. Pain is usually accentuated by exposure of the hand to cold. Early diagnosis and prompt institution of treatment are mandatory to prevent the disastrous consequences of "fixed" sympathetic dystrophy. Early institution of repeated stellate ganglion blocks, with or without the use of oral sympathetic blocking agents, gradually relieves many patients. Sympathectomy may be required in some cases. The earlier appropriate treatment is started, the less chance there is of irreversible joint contractures and permanent loss of function.

Chapter III

INFECTIONS

SOME INFECTIONS OF the hand follow known penetrating injuries. In other cases the mechanism of the introduction of the infection is not apparent. Antibiotics have substantially reduced the incidence of hand infections, but the need for prompt and adequate surgical drainage of abscesses when they occur has not changed.

FELON AND PARONYCHIA

The subcutaneous tissue overlying the volar side of the distal phalanx is known as the terminal pulp space. This area is sealed off from the more proximal parts of the finger. An abscess in the pulp space is known as a felon (Fig. 46). Because the overlying skin is quite inelastic, the development of a felon is rapidly followed by insistent throbbing pain which literally demands prompt surgical drainage.

A paronychia is a pocket of pus along the border of a nail. Although this is less painful than a felon, adequate drainage must be instituted to prevent spread of the infection.

TENDON SHEATH AND PALMAR INFECTIONS

Infections of the tendon sheaths and palmar spaces are serious problems. Unless the infection is eradicated promptly, the tendons and the surrounding structures will become bound down in dense scar tissue. The resultant stiffness and loss of function is a very difficult problem to overcome.

Infections of the tendon sheaths and palmar spaces may be less painful than a felon at first, because the infection may spread along the tendon sheaths and palmar spaces rather than produce

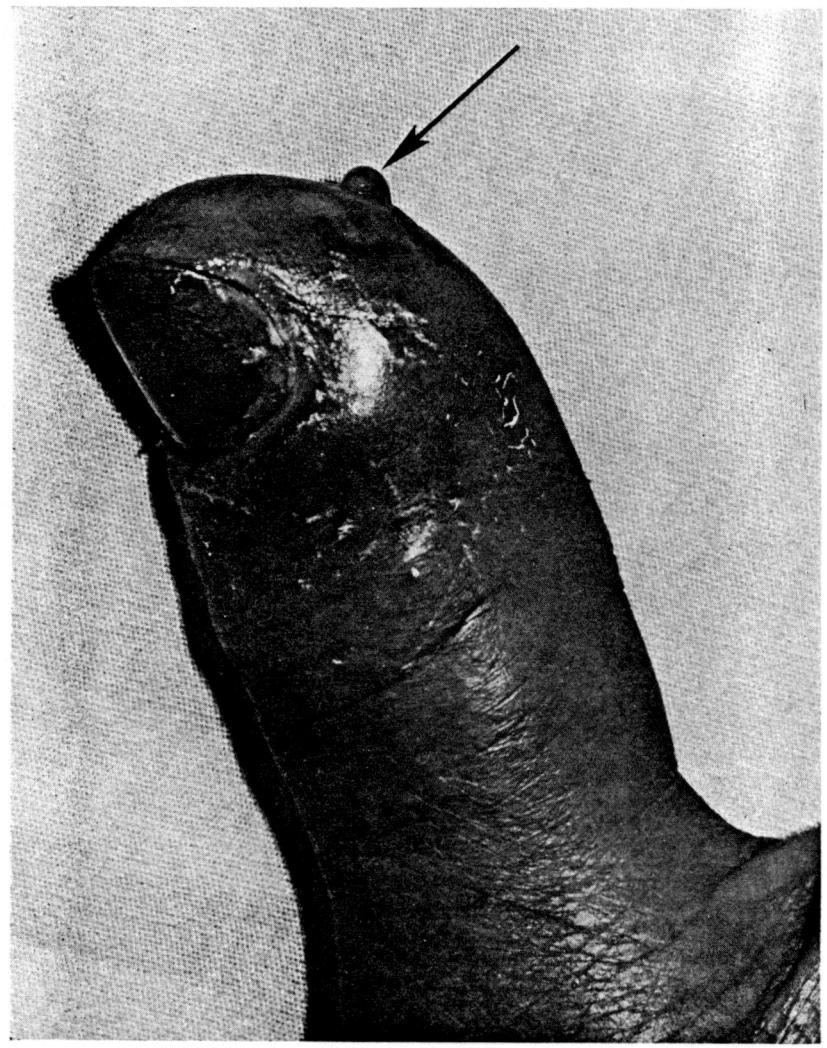

Figure 46. Felon of the thumb. Note the tight swelling of the pulp space and the drop of pus (arrow) which is draining spontaneously.

a painful localized abscess. Early diagnosis is extremely important since a delay of even a day or two may result in severe damage to the hand. Swelling, erythema, and tenderness over a tendon sheath or in the palm, with or without a history of penetrating injury, strongly suggest infection (Fig. 47). Systemic signs are usually minimal in early cases.

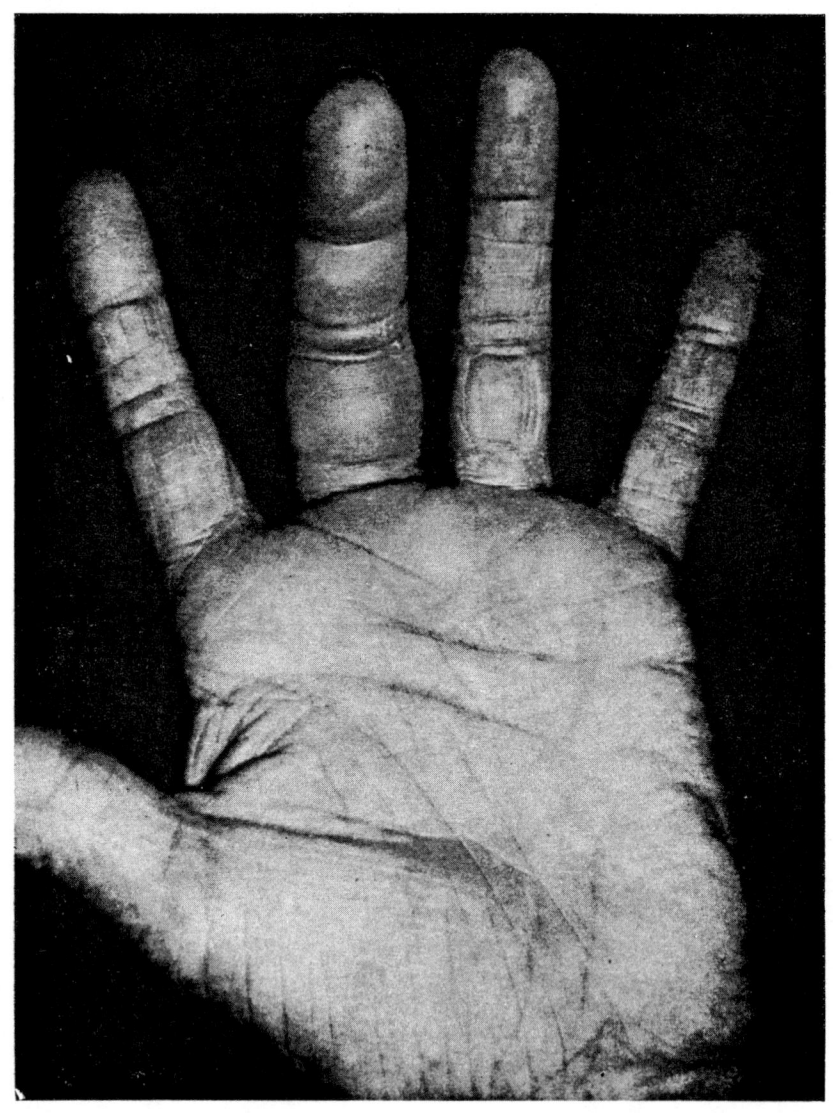

Figure 47. Infection of the flexor sheath of the long finger. Note the generalized swelling of the finger, most marked in the proximal one-third. Pain prevents full extension of the finger, even though the other fingers are in extension.

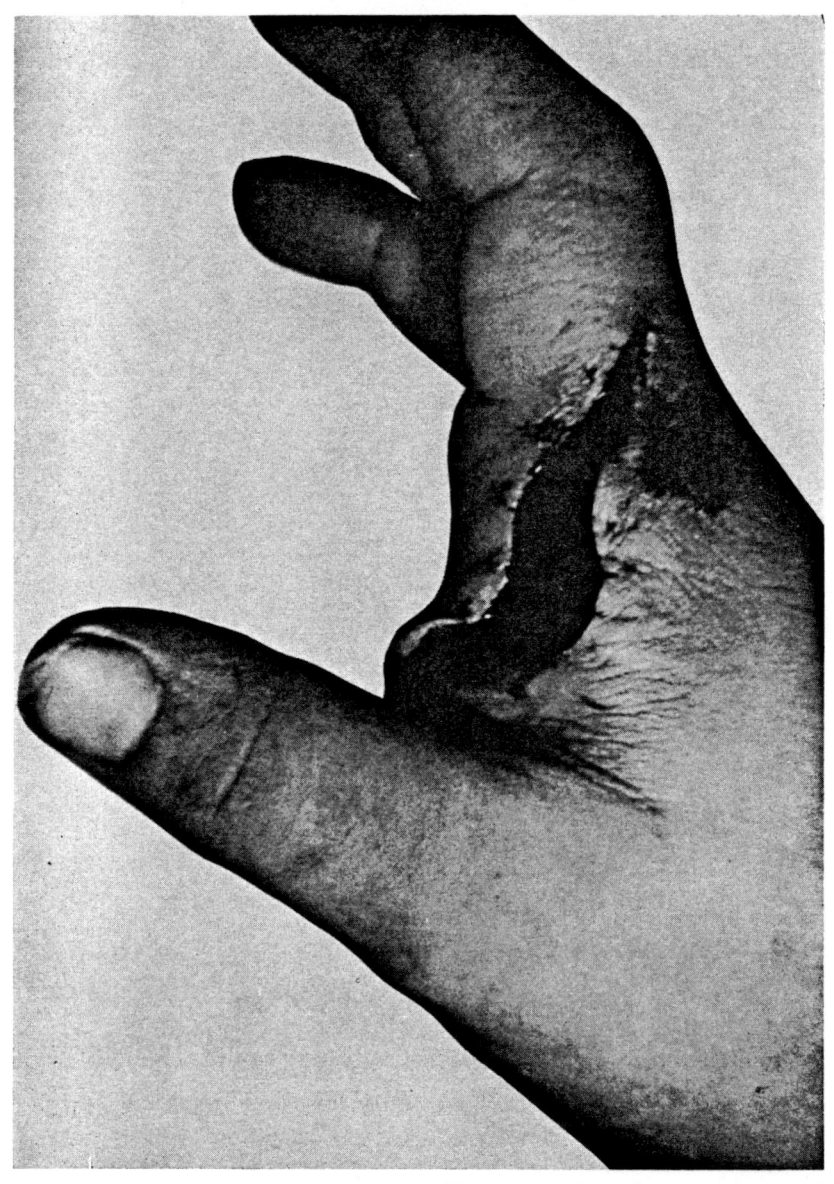

Figure 48. Thenar space infection following a dirty lacerating wound which was sutured. Note the widely abducted position of the thumb.

A thenar space infection in a four-year-old child is illustrated in Figure 48. Four days earlier the child had sustained a laceration of the first web, which was sutured. At the time this photograph was taken, the child had no systemic symptoms. The rectal temperature was 100 degrees, and the white blood cell count was 8,000. At surgery, nevertheless, there was pus covering the dorsal and volar surfaces of the adductor pollicis muscle, and the pus enveloped the flexor tendon sheaths and the sensory nerves to the index and the thumb. The abscess was opened widely and packed. Two days later a loose secondary closure was performed. Three weeks subsequent to the initial surgery, the child had made a full recovery.

Often an abscess seemingly localized to the palm has a small channel extending between the metacarpals and communicating with a pocket of pus on the dorsum of the hand. This condition is known as a collar-button abscess because of its configuration. Drainage of the obvious palmar abscess is insufficient; the dorsal abscess must be drained as well. Delay in recognition of the dorsal component prolongs the infection and greatly increases the likelihood of permanent damage to the hand. In most cases there will be an area of localized swelling and erythema on the dorsum of the hand.

HUMAN BITES

Human bites are often followed by severe infections, probably because there is a sociological correlation between the personality of biters and poor oral hygiene. Every human bite, no matter how superficial, must be treated as a *major injury*. After thorough cleansing, pack the wound open. *Do not suture it.* Immobilize the hand and put it at rest. Start the patient on clindomycin HCl hydrate, 150 to 300 mg every six hours. Check the patient and the wound daily. If there is no suggestion of infection on the third or fourth day, a large wound may be loosely closed with a few sutures over a drain.

Early diagnosis and prompt, adequate surgical drainage are essential to the successful treatment of pyogenetic infections of the hand. The role of antibiotics in hand infections, although important, is secondary.

Chapter IV

DEGENERATIVE DISEASES

THE PRIMARY FUNCTIONS of the hand are pinching and grasping. Small objects are usually held between the tips of the thumb and one or more of the fingers; this is called *pinch*. Large objects are normally held against the palm by the thumb and all four fingers; this method of holding is designated *grasp*. Joints, muscle-tendon units, and sensation all play key roles in effective pinch and grasp.

Mobile, pain-free joints are necessary to enable the patient to position the thumb and fingers in the proper relationship. The joints must also be stable so that their positions are maintained when the muscles are contracted and power is applied. The joints of the thumb and fingers are those primarily involved in pinch and grasp. For strong pinch and grasp, the integrity of the radiocarpal, intercarpal, and carpometacarpal joints is also necessary.

The power of grasp is maximal when the wrist is fully extended. As the thumb and fingers are flexed, the full power of the thumb and finger flexors is utilized if the wrist is in full extension. (Make a tight fist. Note that your wrist has automatically gone into extension. Now flex your wrist while continuing to make a tight fist. Note how much weaker your grip is now.)

Enough functioning muscle-tendon units must be available to position, stabilize, and move the involved joints. Many of the joints of the hand are not simple hinge joints and are therefore capable of motion in multiple planes. Thus, a large number of muscle-tendon units is needed to stabilize and move these joints. When pinching an object between thumb and index, for example, the coordinated use of the following muscle-tendon units is involved:

1. The eight extrinsic flexors and extensors of the thumb and index.

2. The four thumb intrinsic muscles of the thenar group.

3. The volar and dorsal interosseous muscles of the index finger.

4. The wrist flexors and extensors which stabilize the wrist. The loss of just a few of these muscle-tendon units may produce significant disability of the hand.

Normal sensation is essential to the precise and rapid function of the hand. We direct our hands through various sensory modalities, especially proprioception and the various forms of touch. Although a hand with deficient sensation can be controlled visually, as we operate a pair of pliers, visual control results in slower and less precise use of the hand.

Sensation also plays an important role in protecting the hand from injury. Perception of painful stimuli often enables the patient to avoid or to minimize injury to the hand. However, the quality of sensation necessary to protect the hand from injury is more primitive than that necessary for precise and rapid function of the hand. This cruder quality of sensation is termed *protective sensation*. A good example of the need for protective sensation is the patient who has anesthesia from a median nerve injury. He frequently burns himself when holding a cigarette between his fingers.

Most people are unaware of the almost incalculable number of movements which their hands perform each day. Consider the number of hand motions made by a typist during an eight-hour day. Children and young adults generally tolerate the use to which their hands are subjected. However, some middle-aged individuals seem to be prone to develop degenerative changes about the hands. A constitutional predisposition may be involved, as many of the same patients are prone to develop degenerative conditions involving the fibro-osseous systems elsewhere in the body, such as in the shoulder and low back. The most vulnerable structures of the hand are the joints, the tendons, and the tendon sheaths.

The tendons and their sheaths are subject to several degenerative conditions which are distinct clinical entities. The commonest of these are tenosynovitis, calcific tendinitis, and trigger finger and trigger thumb. They are all observed most frequently in middle-aged women.

TENOSYNOVITIS

Tenosynovitis is an inflammation of a tendon sheath. Although tenosynovitis can be caused by infection (see "Infections"), it is usually a localized degenerative phenomenon. In some cases it is related to occupation. The most common site for tenosynovitis in the hand is at the radial styloid process, where the sheaths of the extensor pollicis brevis and the abductor pollicis longus are thick-walled, inelastic, and firmly attached to the underlying bone. In this location the condition is known as de Quervain's disease, and here tenosynovitis is rarely, if ever, due to infection. Involvement may be bilateral. Pain with use of the thumb is the chief complaint. The patient often complains of weakness, as the pain limits the power of pinch and causes the patient to drop objects. A hard, tender-pointed swelling is sometimes observed over the radial styloid. X-rays are negative for osseous changes. The diagnosis is made by the reproduction of pain when the examiner simultaneously deviates the wrist ulnarward and pulls the thumb down across the palm into maximum adduction and flexion. This maneuver is known as the Finklestein test and is illustrated in Figure 49. Mild cases of recent onset will often respond to rest or to the injection of dexamethasone into or about the inflamed tendon sheath. More advanced cases will require partial excision of the thickened tendon sheath.

CALCIFIC TENDINITIS

Calcium deposition can occur in focal areas of degeneration in tendons. The condition is designated *calcific tendinitis*. In the hand this condition occurs most frequently in the tendon of the flexor carpi ulnaris, near its insertion into the pisiform. In addition to pain and swelling, erythema may overlie the tendon, suggesting the possibility of infection. However, visualization of

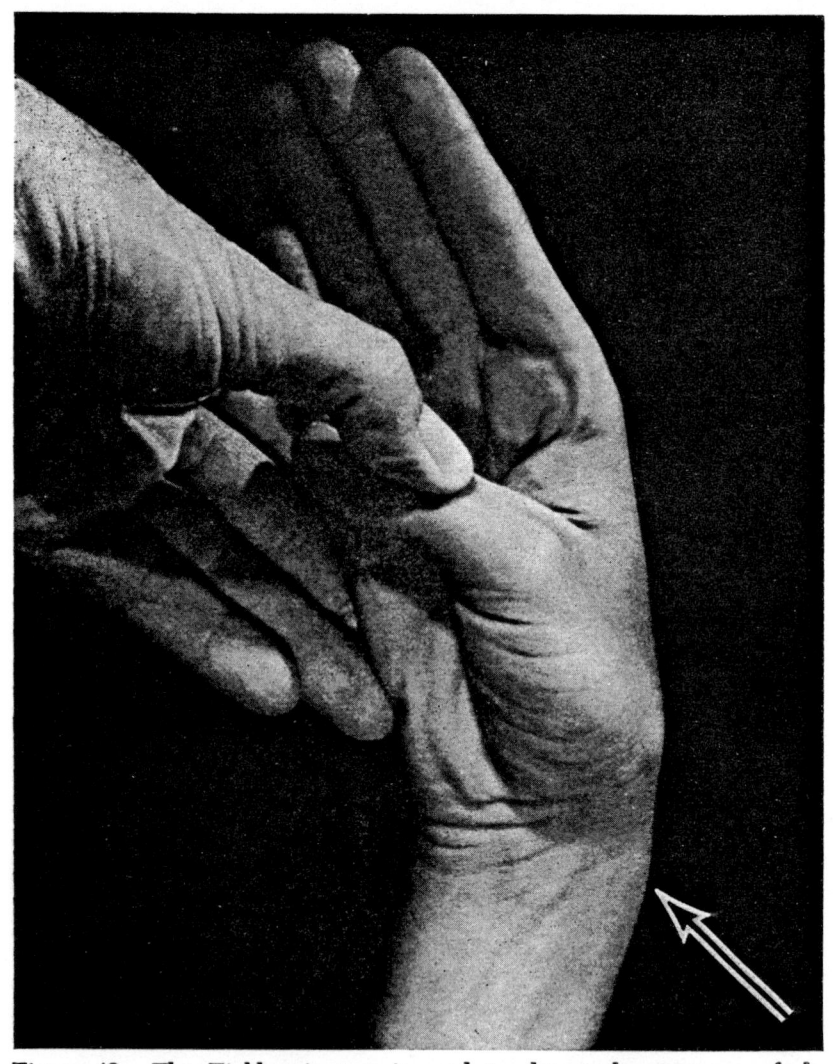

Figure 49. The Finklestein test is used to detect the presence of de Quervain's disease. Pain is reproduced at the radial styloid (arrow) when the examiner holds the wrist in ulnar deviation and simultaneously flexes and adducts the thumb.

calcific deposits in the tendon on x-ray confirms the diagnosis (Fig. 50). The symptoms in most cases will respond to rest, superficial x-ray therapy, or local injection of dexamethasone. When injecting the area, one must be careful not to injure the

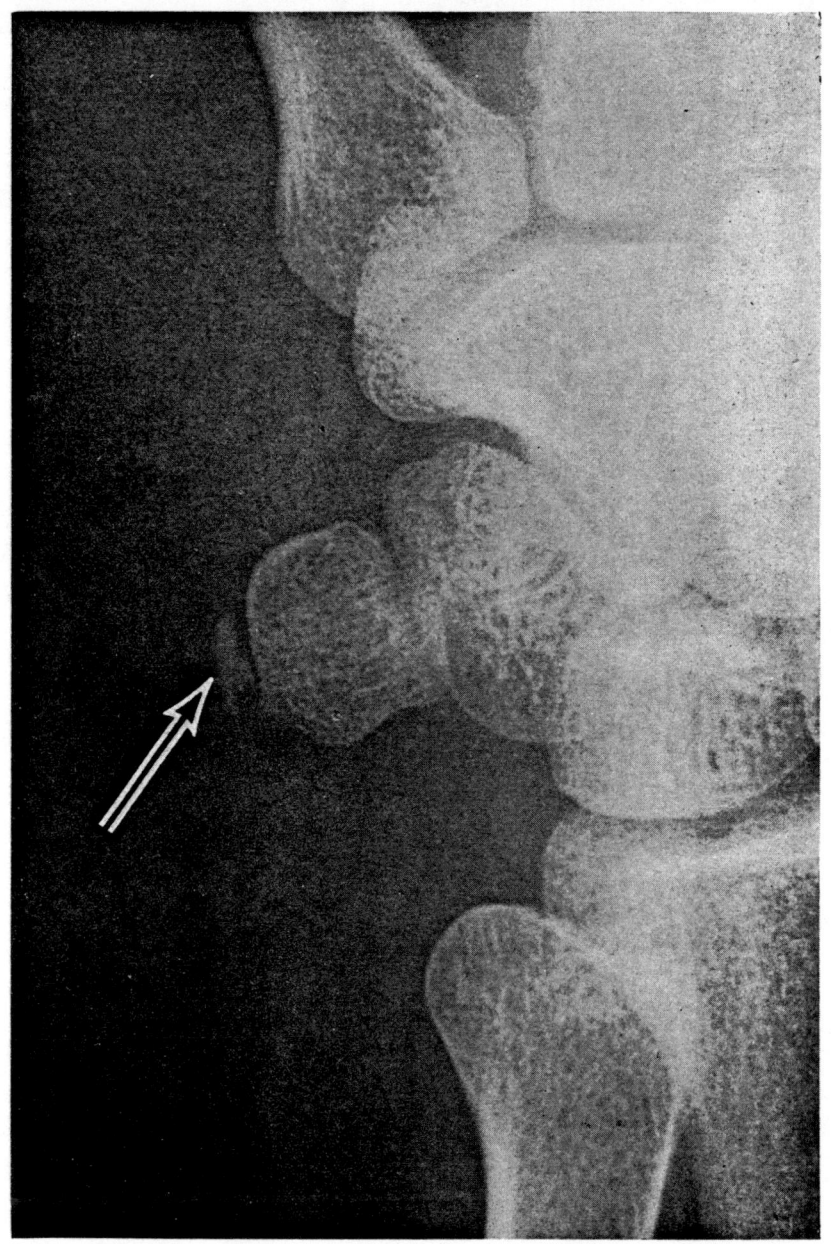

Figure 50. Calcific tendinitis (arrow) of the flexor carpi ulnaris.

ulnar nerve or vessels which lie directly beneath the flexor carpi ulnaris at this level. Occasionally surgery may become necessary if pain is severe or persists despite conservative treatment.

TRIGGERING

The profundis and sublimis tendons of each finger, as well as the flexor pollicis longus, pass through separate tendon sheaths. Beginning in the distal palm and extending out to the distal phalanges, these sheaths are inelastic and snug-fitting. They are especially snug-fitting in the region of the MP joints. Middle-aged and older people are prone to the development of fusiform enlargement of one or more of these tendons at the level of the MP joints. This condition, when symptomatic, is called *trigger finger* or *trigger thumb*.

Initially, the patient is aware of a snapping sensation as he extends the digit from the flexed position. The sensation is most noticeable on arising in the morning. The snapping may be painful, and it may occur both when the patient extends and when he flexes the involved digit. In advanced cases the patient has more difficulty in extending the digit than in flexing it, because the extensor muscles are not as strong as the flexor muscles. As a consequence, he may be able to flex the digit actively but be unable to extend the digit except passively with the help of the other hand. In extreme cases the digit becomes locked. The locked position may be either flexion or extension. Local injections of dexamethasone will often relieve mild cases which have been symptomatic only a few weeks. In more advanced cases it will be necessary to open the tendon sheath in the involved area to allow the enlarged section of the tendon to glide freely.

OSTEOARTHRITIS

Osteoarthritis is the term used for degenerative changes involving a joint. In most cases these changes occur with the normal wear and tear associated with the passage of years. In others the condition develops following direct trauma to a joint, and occasionally the disease appears to be due to an acceleration of the aging process somehow localized to specific joints.

Regardless of the pathogenesis, the symptoms of advanced osteoarthritis involving the joints of the hand are pain, instability, and/or loss of motion. It is unusual for more than one or two

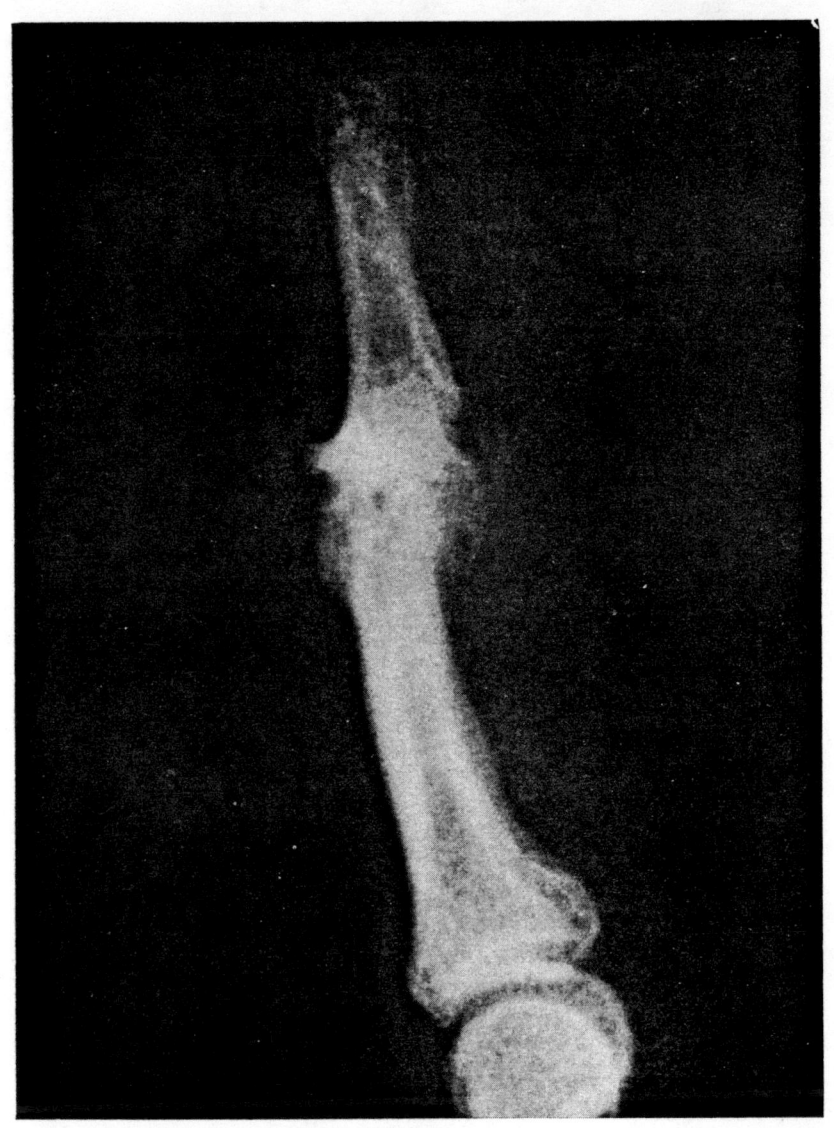

Figure 51. Advanced osteoarthritis of the DIP joint. Note the spur formation and loss of articular cartilage.

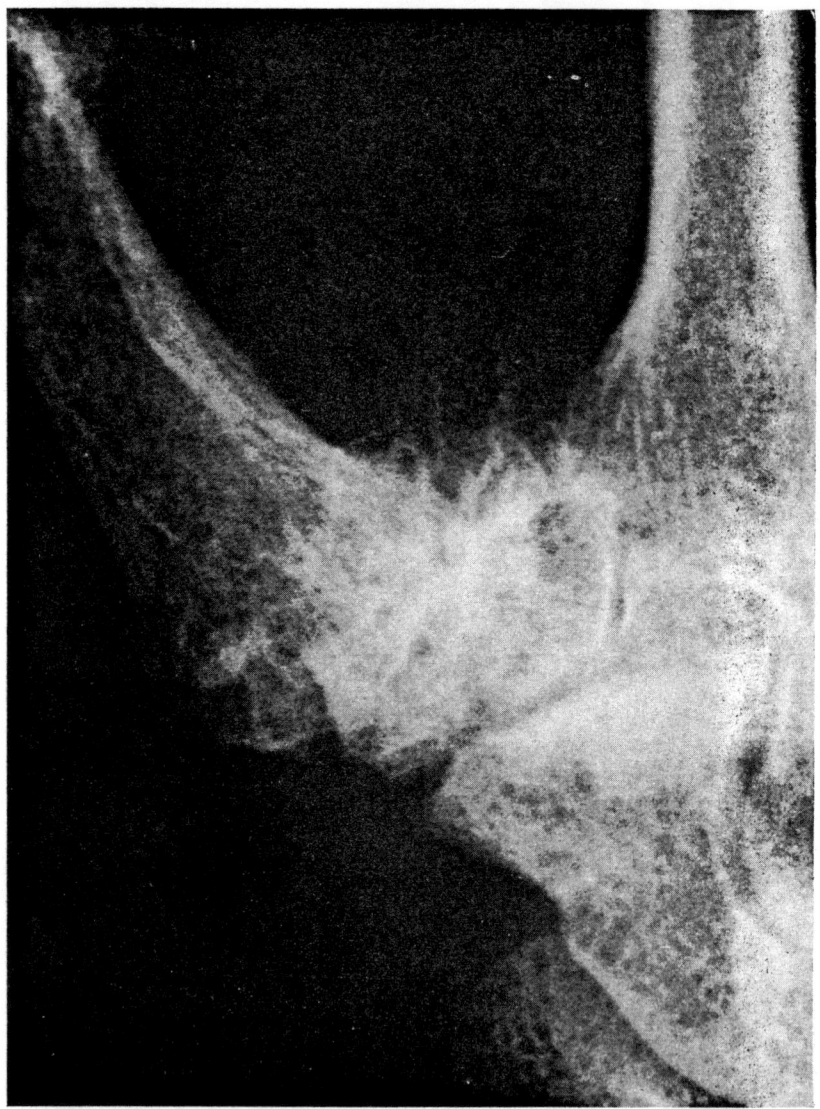

Figure 52. Advanced osteoarthritis of the first carpometacarpal joint.

joints to deteriorate to the point of requiring surgery, and those joints most often involved are the DIP (Fig. 51), the 1st carpometacarpal (Fig. 52), and the joints of the wrist (Fig. 53). However, the function of the entire hand can be adversely affected by a single painful or unstable joint (Figs. 54 and 55).

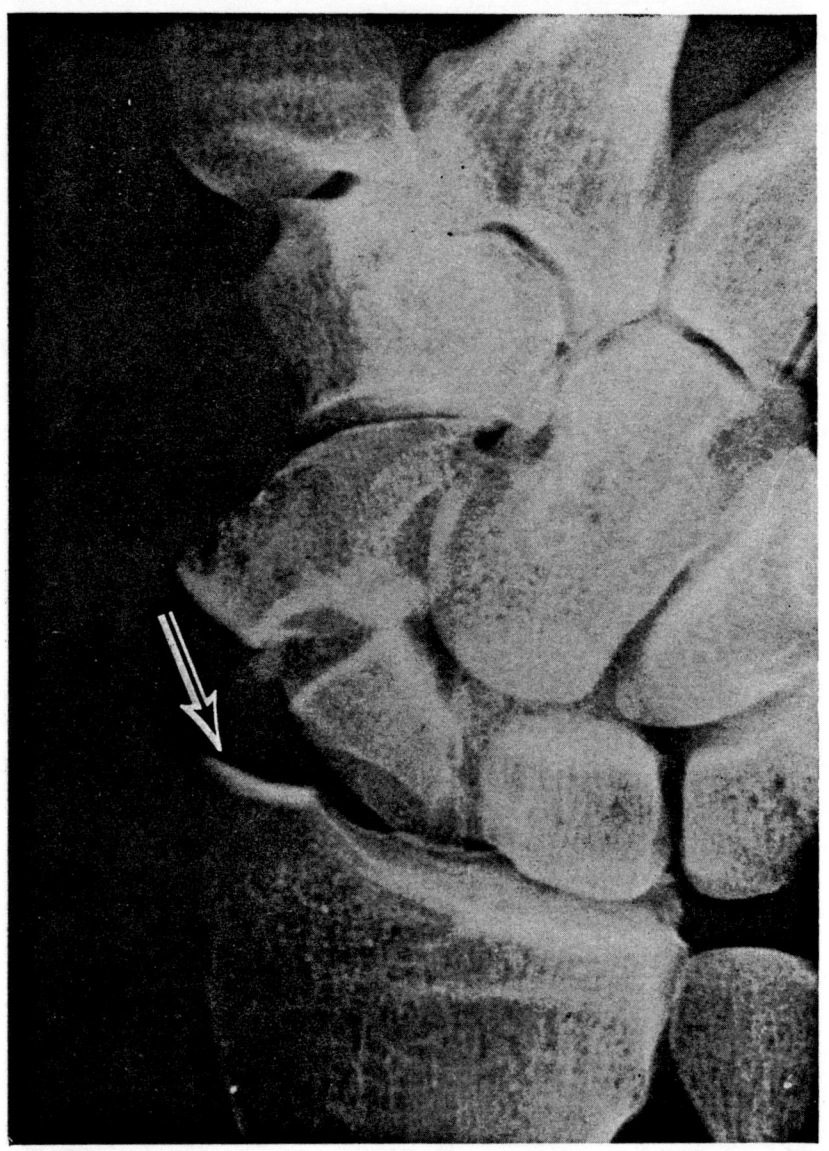

Figure 53. Osteoarthritis of the wrist secondary to nonunion of the navicular. Note the dense white sclerosis of the deformed radial styloid (arrow).

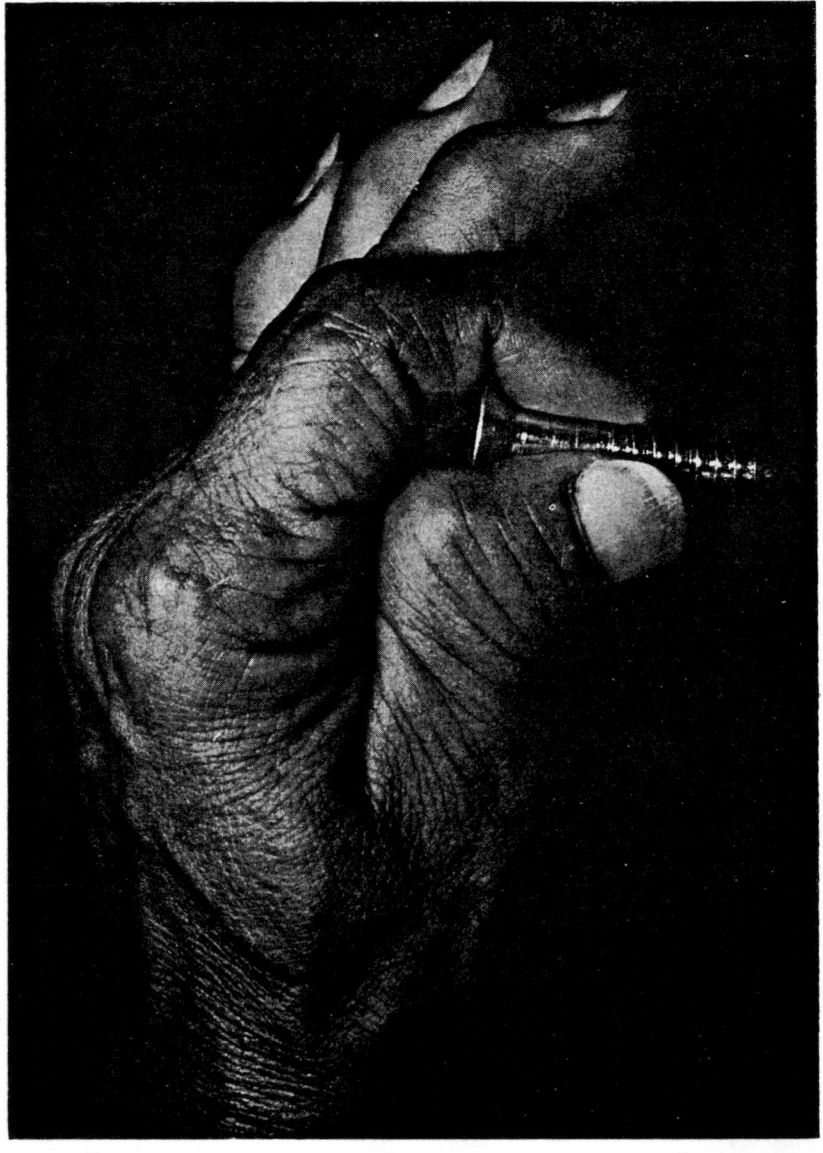

Figure 54. Extreme arthritic deformity of the thumb IP joint. The joint is grossly unstable, with 70° of abnormal lateral angulation.

Degenerative Diseases 77

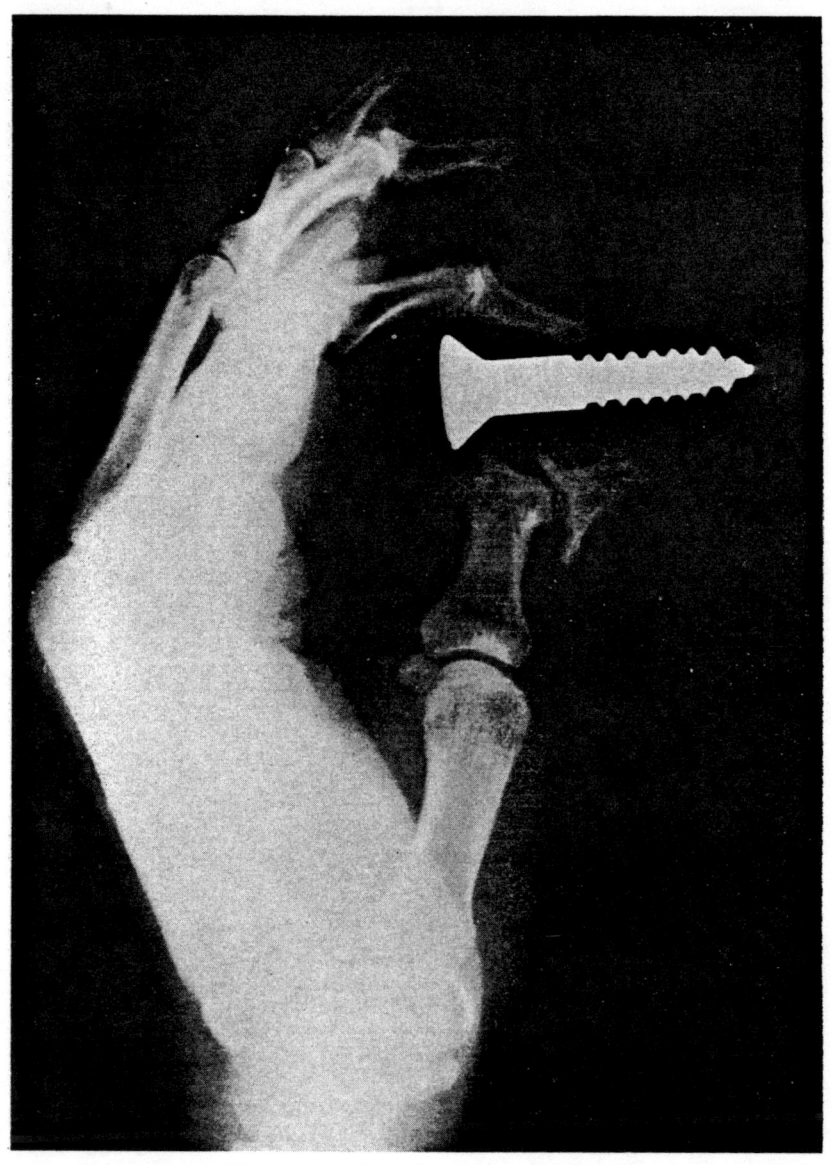

Figure 55. X-ray of the hand shown in Figure 54.

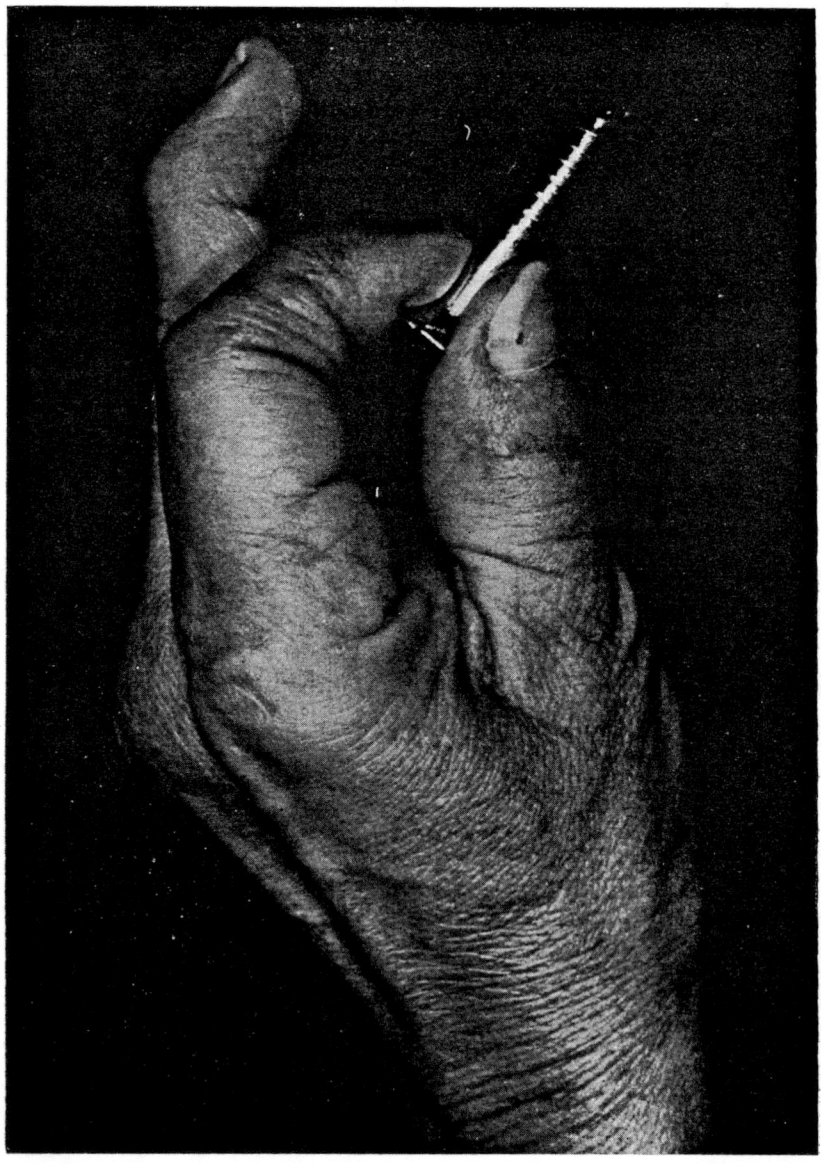

Figure 56. Hand of the patient seen in Figure 54 after arthrodesis of the thumb IP joint.

There is no treatment known which will alter or slow the course of osteoarthritis, and it is impossible at the onset to predict how rapidly the condition will progress or how severely a joint will be affected. Early in the disease various medications and other modalities, such as heat, will provide palliative relief of pain and stiffness, and surgery is not warranted at this time. Later, when a joint has become severely disabled, surgery is the only solution. The most important surgical consideration is to provide relief of pain and stability of the joint. A stiff joint which is, however, painless is less handicapping than one which retains motion but is painful and/or unstable. Furthermore, by the time surgery is contemplated, the patient has frequently lost some motion. Therefore, a procedure designed to eliminate pain and stabilize the joint, although sacrificing all motion in the process, is the treatment of choice. Such an operation is arthrodesis. The cartilage and subchondral bone of the joint surfaces are excised, and the raw ends of the two bones are brought into contact and immobilized to induce solid union of the bones. The net effect is that pain and instability have been eliminated, but all motion in the joint has been lost. However, the stiffness of the joint is more than compensated for by the increased strength and stability gained (Fig. 56).

An alternative surgical approach is arthroplasty, which attempts to reconstruct the joint. The end results are more variable and less predictable than those from arthrodesis, and it is only rarely that substantial pain relief with both stability and an adequate range of motion will be obtained. In the hand, arthroplasty is primarily used for the MP joints of the fingers, as complete loss of motion in these joints is very disabling.

Hard and fast rules cannot be made for a decision between arthrodesis and arthroplasty. Many factors must be weighed, including the number and the specific joints actually or potentially involved; age, sex, and occupation of the patient; and the patient's preference.

The use of silicone motion-preserving implants is still experimental in the treatment of osteoarthritis and cannot be recommended at this time.

CARPAL TUNNEL SYNDROME

Carpal tunnel syndrome is the name given to neuropathy of the median nerve caused by compression at the wrist. Carpal tunnel syndrome may be unilateral or bilateral. It is most frequently seen in middle-aged women. A history of trauma or disease involving the wrist is infrequent. However, compression neuropathy of the median nerve at the wrist occasionally results from trauma, such as Colles' fracture. In such cases, x-rays will usually demonstrate a fracture fragment encroaching on the carpal canal; the median nerve is compressed between the transverse carpal ligament and the flexor tendons. The median nerve may also be compressed in the carpal canal by a mass of hypertrophied synovial tissue in patients suffering from rheumatoid arthritis.

The patient complains of tingling in one or more digits innervated by the median nerve. The paraesthesias may awaken the patient at night. Often the paraesthesias are produced by playing the piano, knitting, or other activities which require repeated flexion of the wrist. The tingling is inconstant at first; then, as the disease progresses, the tingling becomes more marked and may eventually become constant. In advanced cases the patient may also complain of aching in the volar aspect of the lower forearm. The patient rarely complains of weakness of grip.

Examination will usually demonstrate some alteration of sensation in the distribution of the median nerve distal to the wrist. Touch and pin-prick sensation are usually diminished in some portion of the median distribution. Occasionally the examiner will encounter hyperpathia and hyperesthesia, the patient complaining of an annoying increase in the appreciation of tactile stimuli during the examination. Thenar atrophy may occur as a late manifestation of the syndrome. Such atrophy is best observed when the thenar eminence is viewed in profile (Fig. 57). In unilateral cases the difference between the two hands will be obvious. The examiner may elicit local tenderness over the median nerve at the wrist. Percussion over the median nerve at the wrist may produce paraesthesias in the radial side of the palm, in the thumb and/or fingers (Tinel's sign). When the wrist is held in maximum flexion for not less than 60 seconds (Fig. 58)

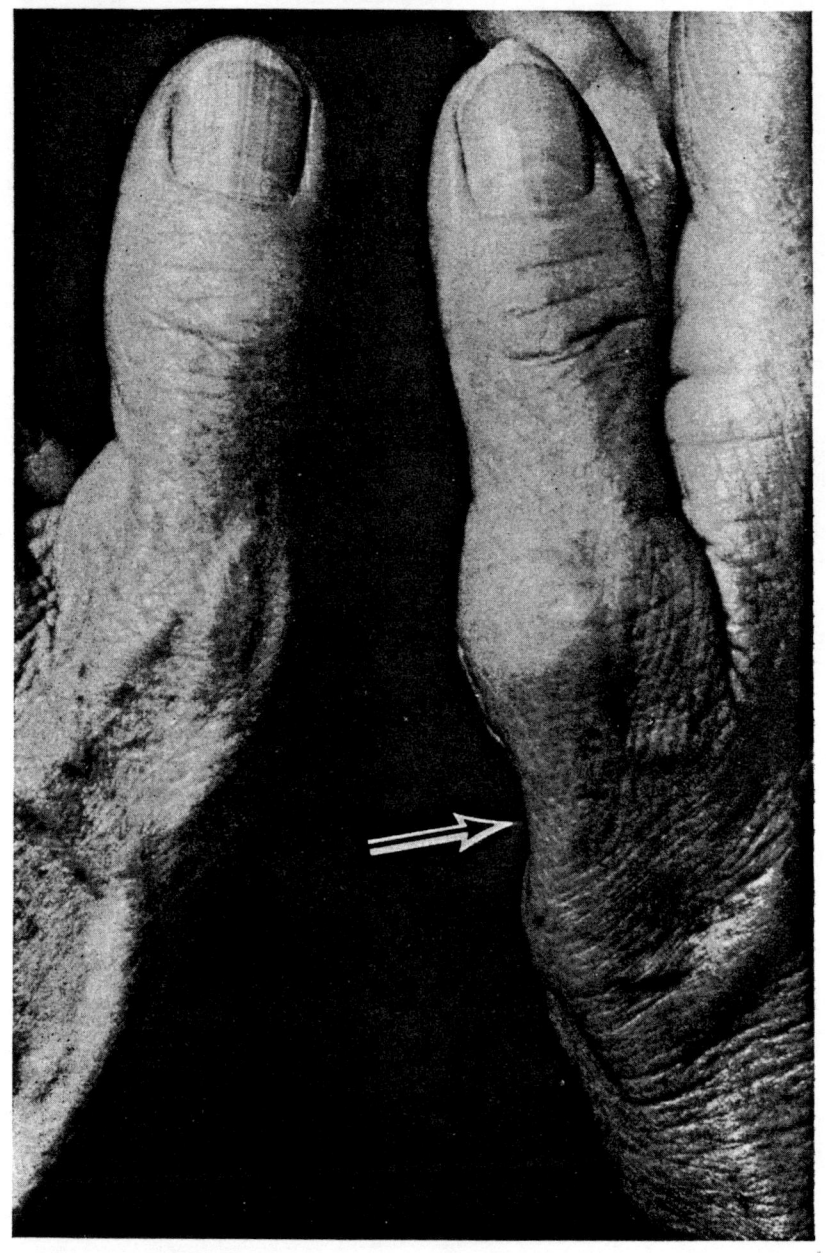

Figure 57. Thenar atrophy is seen best by viewing both thumbs from the dorsal side. The atrophy (arrow) is then seen in profile.

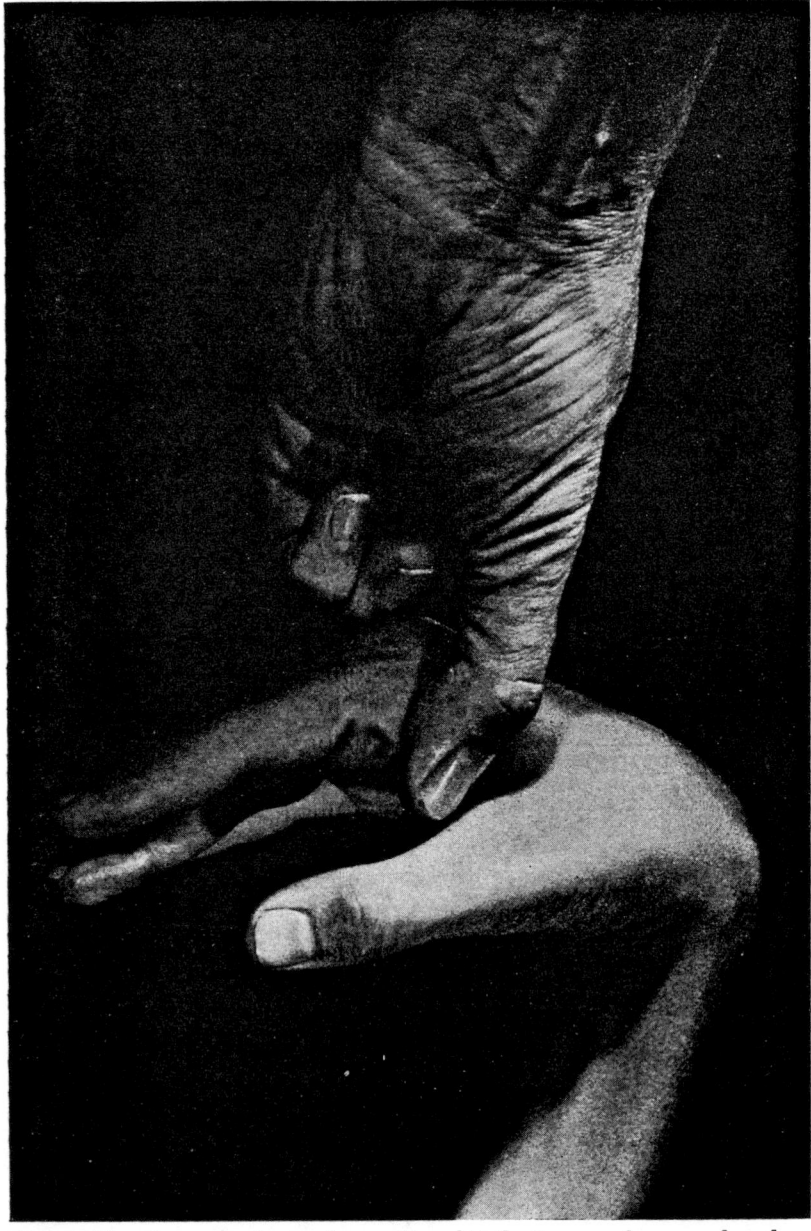

Figure 58. Paraesthesias in the median distribution can be reproduced or accentuated in cases of carpal tunnel syndrome by holding the wrist in maximum flexion for one minute.

and the patient experiences paraesthesias in the distribution of the median nerve (or the paraesthesias already present are accentuated), the diagnosis of carpal tunnel syndrome is established in most patients.

The treatment of carpal tunnel syndrome is surgical decompression of the median nerve at the wrist. The older the patient and the longer the duration of symptoms, the less likely it is that the paraesthesias will be completely relieved. In most cases, however, progressive neurological involvement will be prevented by prompt surgical intervention. Thenar atrophy is not relieved by surgery as often as are paraesthesias.

Although compression neuropathy of the median nerve at the wrist is quite common, the ulnar nerve is rarely involved in a similar fashion. Therefore, when the patient's complaints and/or physical findings implicate the ulnar nerve or the ulnar nerve as well as the median, the physician must look diligently for a cause proximal to the wrist. The pathology may be situated anywhere between the upper forearm and the brain.

Chapter V

RHEUMATOID ARTHRITIS

THE MANAGEMENT OF rheumatoid arthritis in the hand is complicated by the fact that the disease also frequently involves weight-bearing joints of the lower extremities, such as the hips and knees. In such cases, the hands must bear the added burden of using crutches or canes. Therefore, treatment of the rheumatoid hand becomes doubly important.

In the early stages of the disease, hypertrophic synovial tissue locally invades and destroys ligaments, tendons, cartilage, and bone. When minimal x-ray changes indicate that no major structural abnormality has yet developed, the operation of synovectomy may be indicated. This procedure is designed to remove the hypertrophied synovium before it invades and destroys adjacent tissues. The operation seems to produce a *local* remission of the disease in the involved joint which may last for a period of years. Thus, synovectomy performed early in the disease can be a prophylactic operation. If there is full motion of the joint preoperatively, some motion is usually lost following surgery. In other cases, if the mass of synovial tissue is blocking motion, then its removal can enhance joint movement, although restoration of a full range of motion cannot be expected. Synovectomy in the rheumatoid hand is employed principally to prevent destruction of the MP joints and to prevent rupture of tendons about the wrist.

In the late stages of rheumatoid arthritis, x-ray changes reflect the advanced and widespread joint damage (Fig. 59). On examination, the joints are frequently distorted (Figs. 60 and 61), enlarged, swollen, red, and painful, with limited motion and sharply reduced function.

The patient frequently has severe involvement of multiple

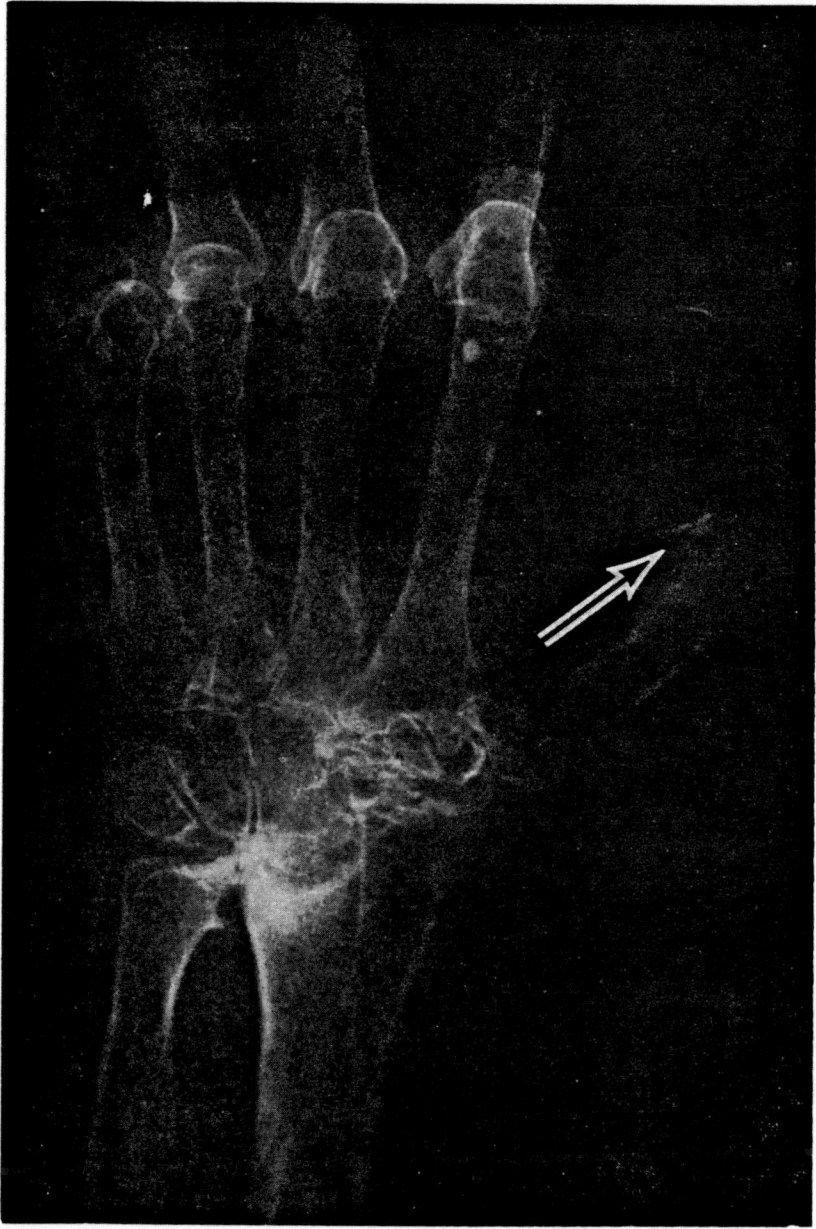

Figure 59. Advanced rheumatoid arthritis. The thumb MP joint is subluxed (arrow). The second, third, and fourth MP joints are completely dislocated. Note the destruction of the carpal bones.

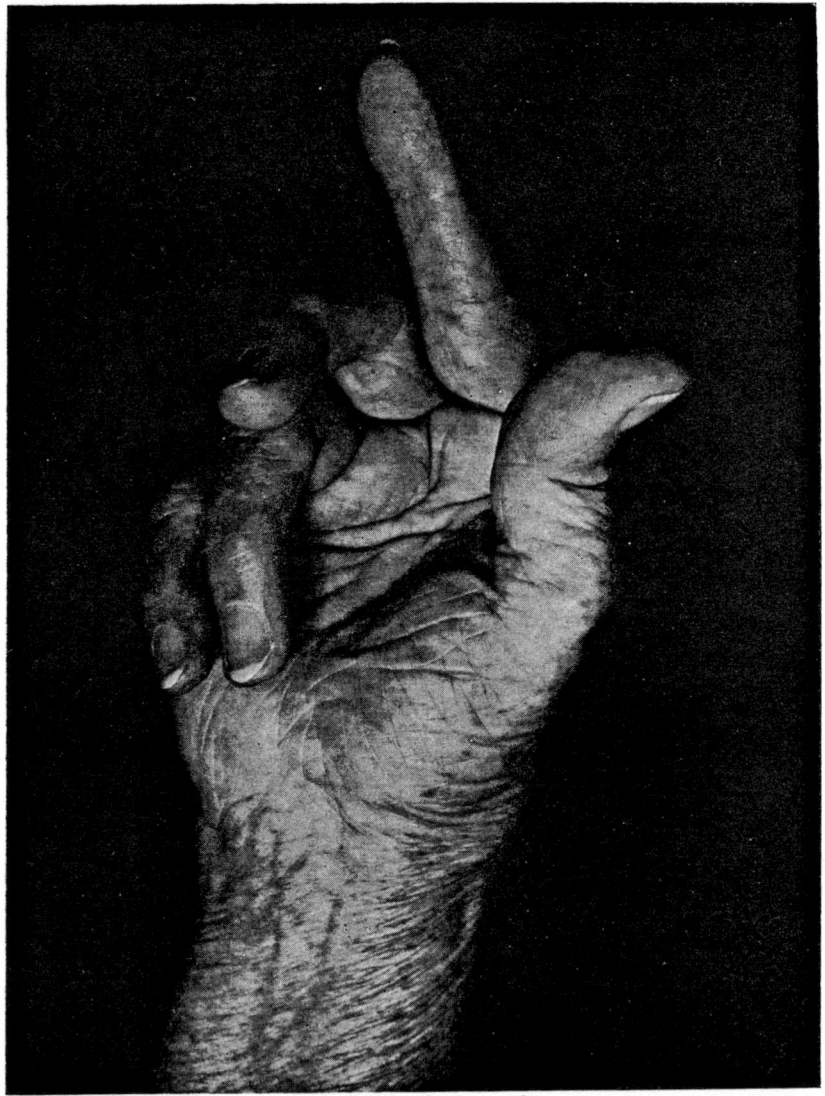

Figure 60. Advanced rheumatoid arthritis. The patient is attempting to extend her fingers fully.

joints of one or both hands—as many as ten or more joints in the same hand, and it is not rare for all three joints of one finger to be severely involved. Surgery at this stage of the disease is a salvage procedure rather than a prophylactic one. The salvage operations are designed to relieve pain, increase function, and improve the appearance of the hand.

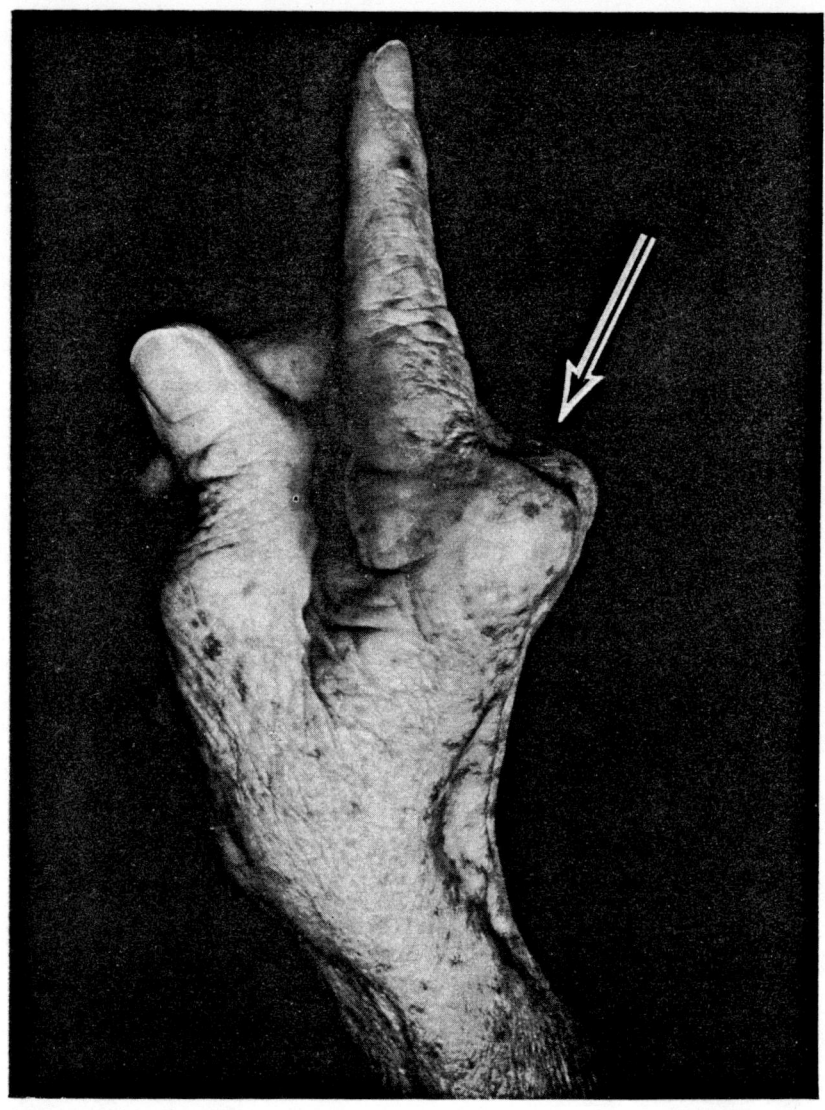

Figure 61. Same case as Figure 60. Note dislocation volarward of the index finger (arrow).

In cases where the MP joints of the fingers are subluxed or frankly dislocated, arthroplasty is usually necessary. In recent years many surgeons have employed silastic implants in the performance of arthroplasties, which has given somewhat better results. At any rate, arthroplasty of the MP joints of the fingers

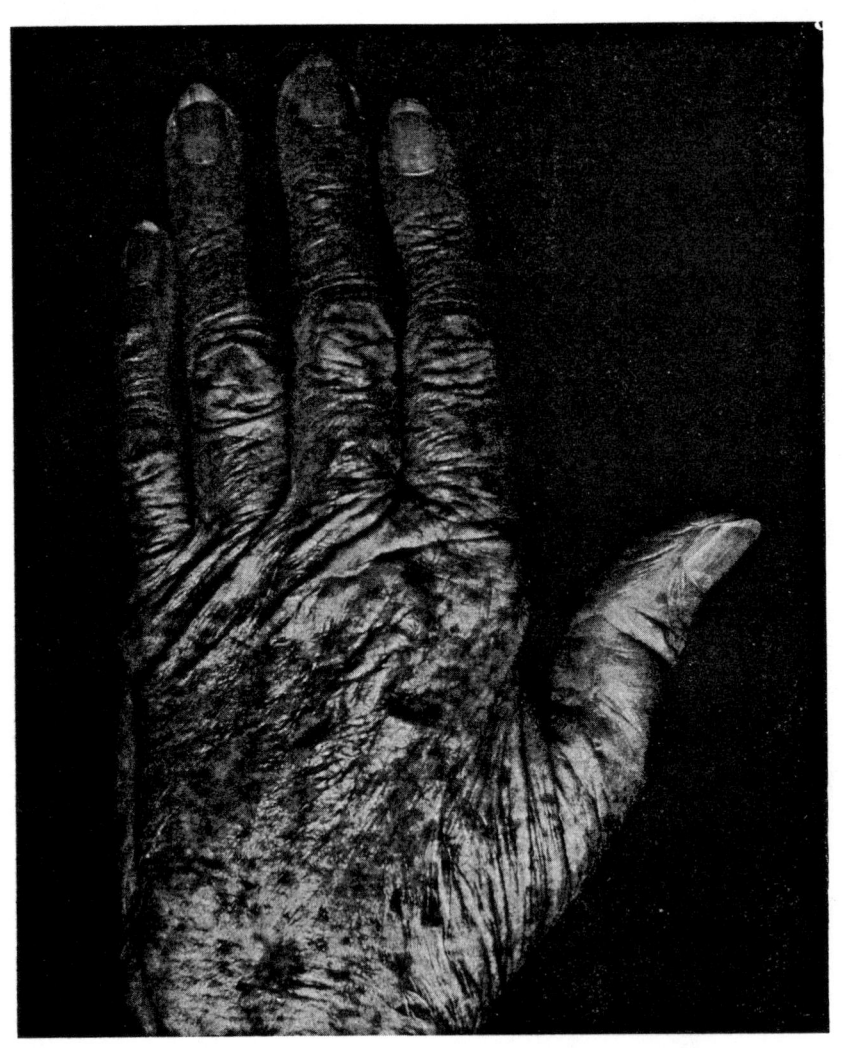

Figure 62. The hand seen in Figures 60 and 61 after arthroplasty of the MP joints of the fingers. She now has full *passive* extension of the joints, but *active* extension is limited.

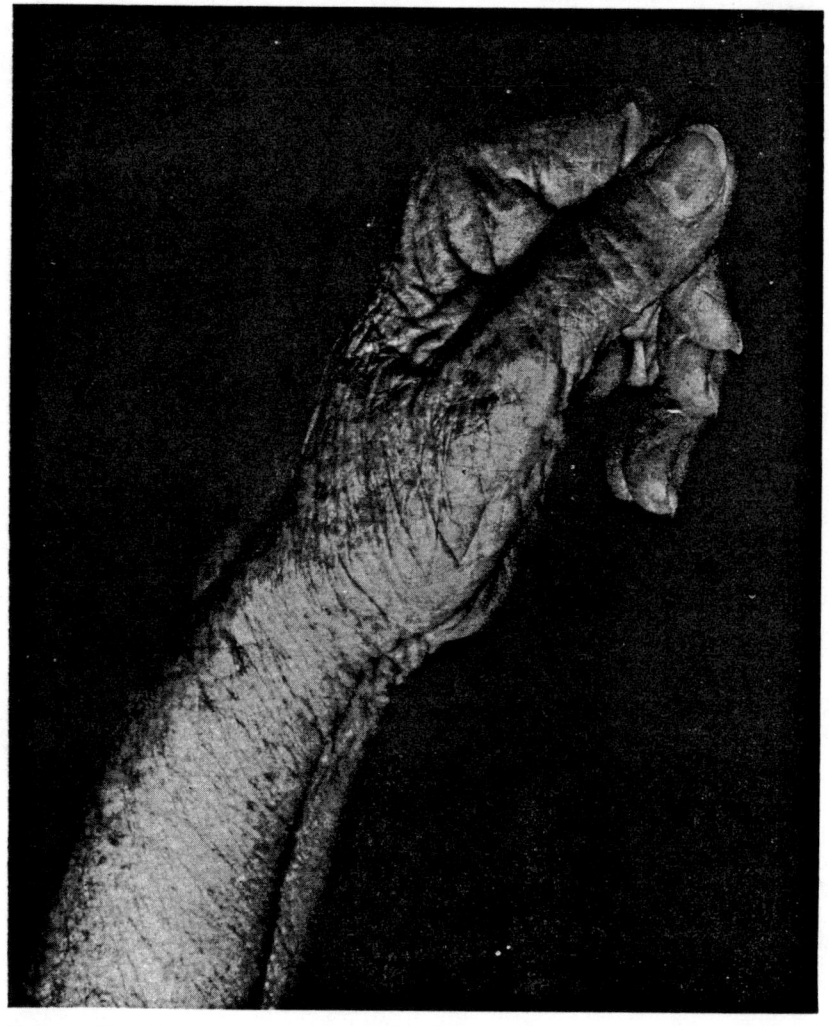

Figure 63. Same case as Figure 62. Flexion of the MP joints is limited to about 45°.

(with or without silicone implants) will decrease or even eliminate pain, and will dramatically improve the appearance of the ugly, dislocated joints (Fig. 62); however, active motion and strength will remain limited (Fig. 63).

Due to the mobility of the *thumb* metacarpal at the carpometacarpal joint, preservation of motion at the thumb MP and IP joints is not essential for good thumb function. Since arthrodesis affords strength and stability as well as relief of pain, it is the operation of choice for the MP and IP joints of the thumb.

In problems involving the PIP joints of the fingers, arthrodesis is also the most reliable operation, since the results of arthroplasty are often unsatisfactory. At the DIP joints, arthrodesis is clearly the procedure of choice because arthroplasty usually results in incomplete relief of pain and little or no motion in these joints.

When rheumatoid arthritis involves either the volar or dorsal aspect of the wrist, tendon ruptures frequently occur after the tendons have been invaded by rheumatoid synovium. This invasion can involve a segment of tendon several inches long, and by the time the tendon finally ruptures, the entire segment has been reduced to a thin attenuated strand of poor quality. For this reason, repair of the rupture is seldom feasible.

The flexor pollicis longus is the tendon most subject to rheumatoid involvement. If this tendon has ruptured there is nothing to balance the pull of the extensor pollicis longus, and the IP joint is gradually pulled into hyperextension, often as much as 90°. This produces the hyperextension deformity of the thumb IP joint so often seen in the rheumatoid hand (Fig. 64). Arthrodesis of the joint is necessary to compensate for the loss of tendon function.

In addition to the tendons about the wrist, the rheumatoid process frequently involves the radiocarpal, intercarpal, and carpometacarpal joints. Pain and swelling are troublesome in the early stages, and gross instability occurs later as cartilage, ligaments, and bone are destroyed. The combination of pain and instability at the wrist severely handicaps function of the hand. Arthrodesis of the wrist usually gives the most satisfactory results. Typically, a bone graft is used to fuse the radius, carpal bones, and the index and long metacarpals together. Excision

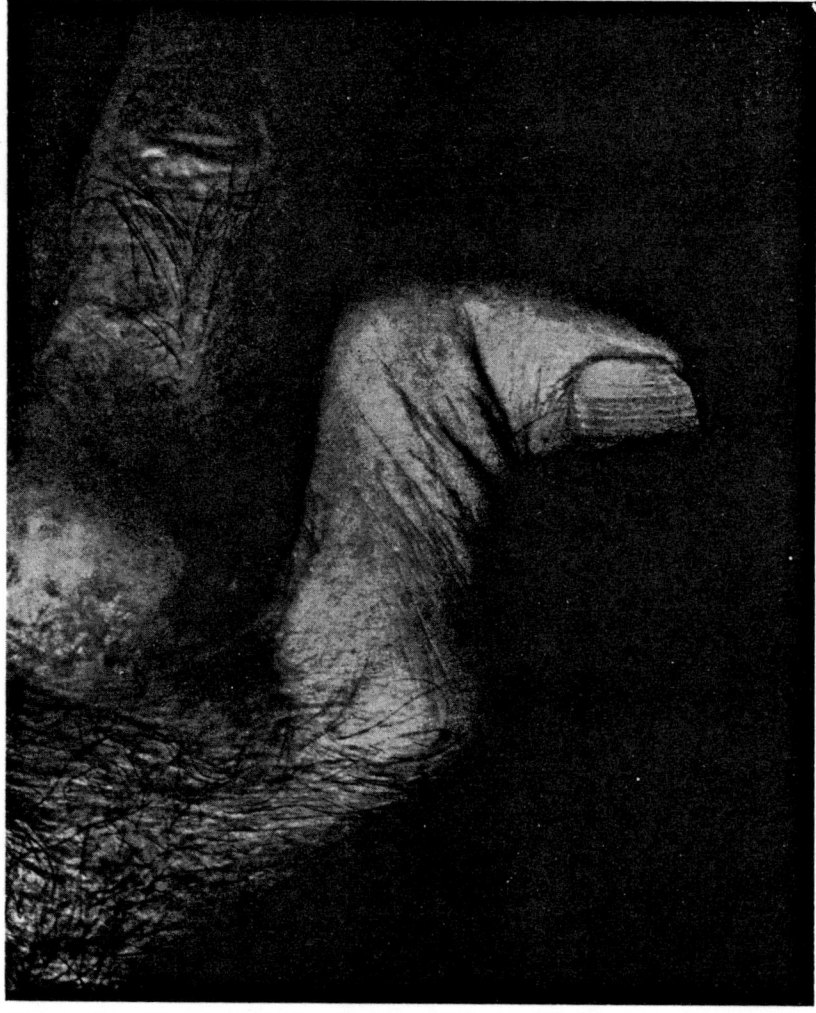

Figure 64. Rupture of the flexor pollicis longus in severe rheumatoid arthritis.

of the distal ulna at the same time may relieve pain associated with pronation and supination, and may substantially increase the range of these two motions. If the extensor tendons are involved, tenosynovectomy can be carried out at the time of the arthrodesis.

The deformities of rheumatoid arthritis tend to develop slowly, and the attending physician does not see rapid change. Consequently, he frequently fails to consider surgical intervention. Then, when the deformities eventually become severe, he may assume that nothing can be done. It is true that surgery cannot restore a rheumatoid hand to normal. Furthermore, in advanced cases where most or all function has been lost, surgery may only provide a modicum of functional improvement. However, Bunnell gave the rationale for surgery in such cases when he said, "When one has nothing, a little is a lot."

The hand in Figure 65, for example, demonstrates total carpal bone destruction with dislocation of the metacarpus off the radius, producing a completely functionless flail hand. Even in this extreme case, arthrodesis (Fig. 66) restored limited function to the thumb and fingers. It is truly unfortunate that the disease was permitted to progress to this advanced stage before surgical intervention was considered.

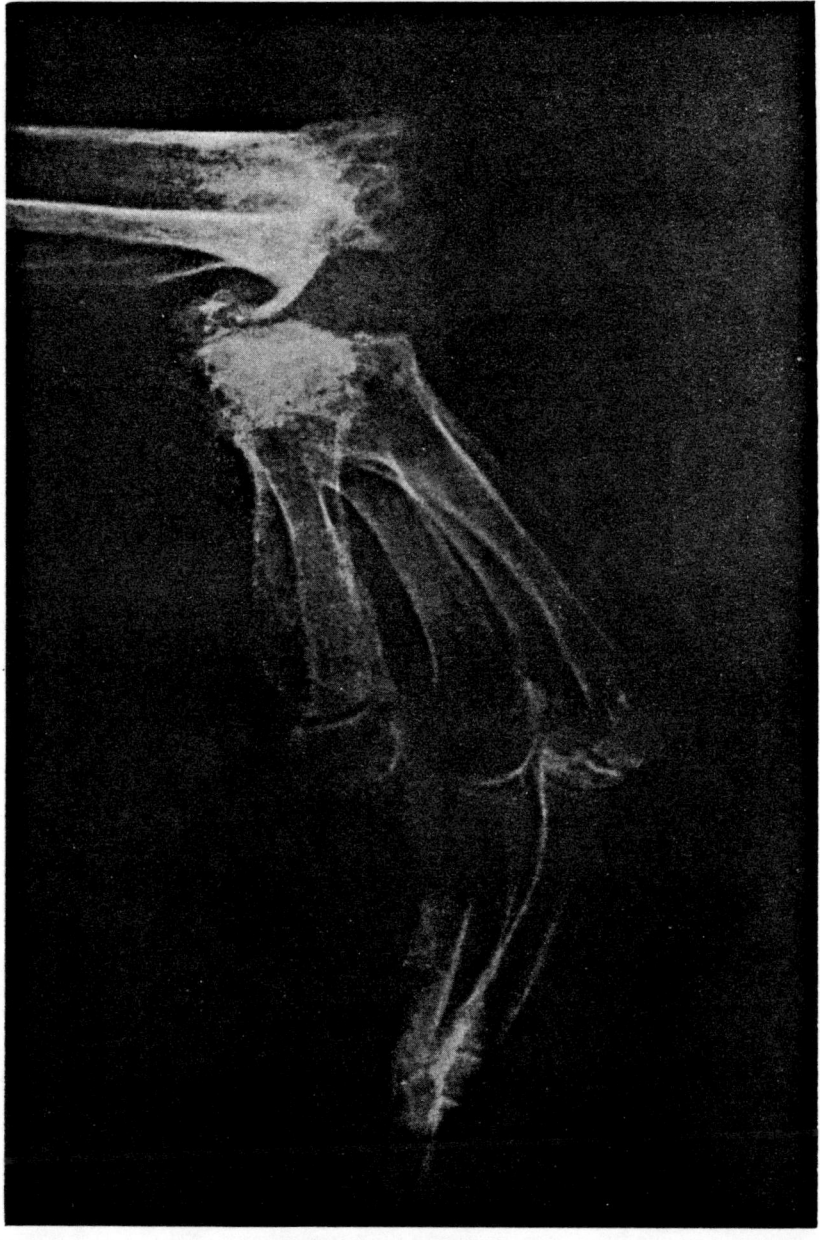

Figure 65. Advanced rheumatoid arthritis has completely destroyed the carpal bones. The hand is functionless.

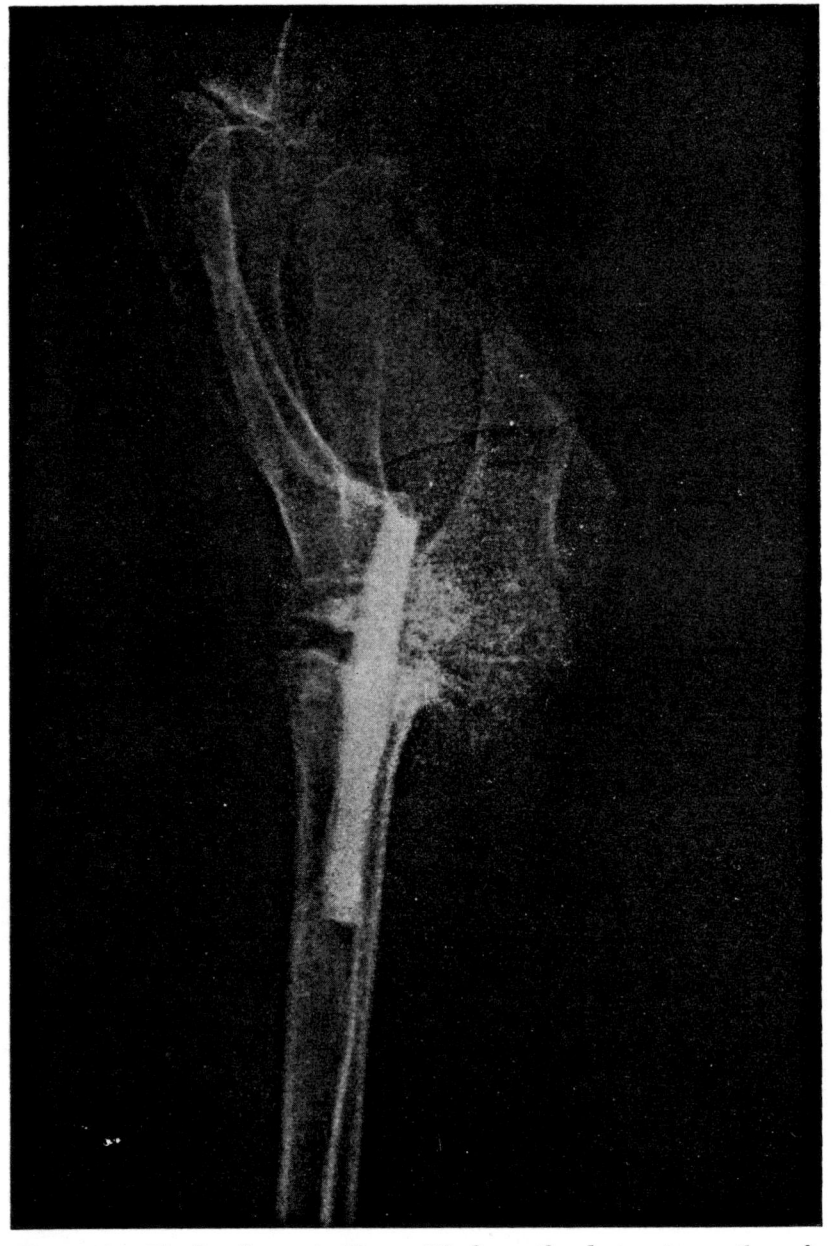

Figure 66. The hand seen in Figure 65 after arthrodesis using a rib graft (dense white).

Chapter VI

DUPUYTREN'S CONTRACTURES

THE PALMAR FASCIA is a tough, thin, inelastic structure on the palmar side of the hand, lying between the thin layer of subcutaneous fat and the underlying nerves, vessels, and tendons. It protects these underlying structures from blunt trauma. The palmar fascia begins in the proximal palm as a continuation of the palmaris longus tendon and extends out over the entire palm, with four distal extensions to the middle phalanx of each finger. These extensions are called pretendinous bands, and they are superficial to the flexor tendons.

Dupuytren's contracture, also known as palmar fibromatosis, is a condition in which areas of hypertrophy develop in the palmar fascia. The first manifestation of the disease is the appearance of one or more firm masses lying just beneath the skin of the palm or of the volar surface of one or more fingers. The hypertrophied fascia tends to push the subcutaneous fat aside and become adherent to the skin, causing dimple-like indentations. The nodules are occasionally tender, especially if they are growing rapidly.

The hypertrophied fascia gradually contracts and draws the involved fingers toward the palm. In some cases the MP joints are affected most (Fig. 67); in others it is the PIP joints (Fig. 68). The ring and little fingers are involved more often than the other digits. The disease occurs most frequently in men of middle age or older, but it is seen in men and women of all ages and is extremely variable in its course. In one patient the disease may progress almost imperceptibly over a period of 20 years. Another patient may develop severe contractures within a year of the time he first notes the appearance of a nodule in his palm. An occasional patient may have rapid progression in one hand and a nearly static condition in the other.

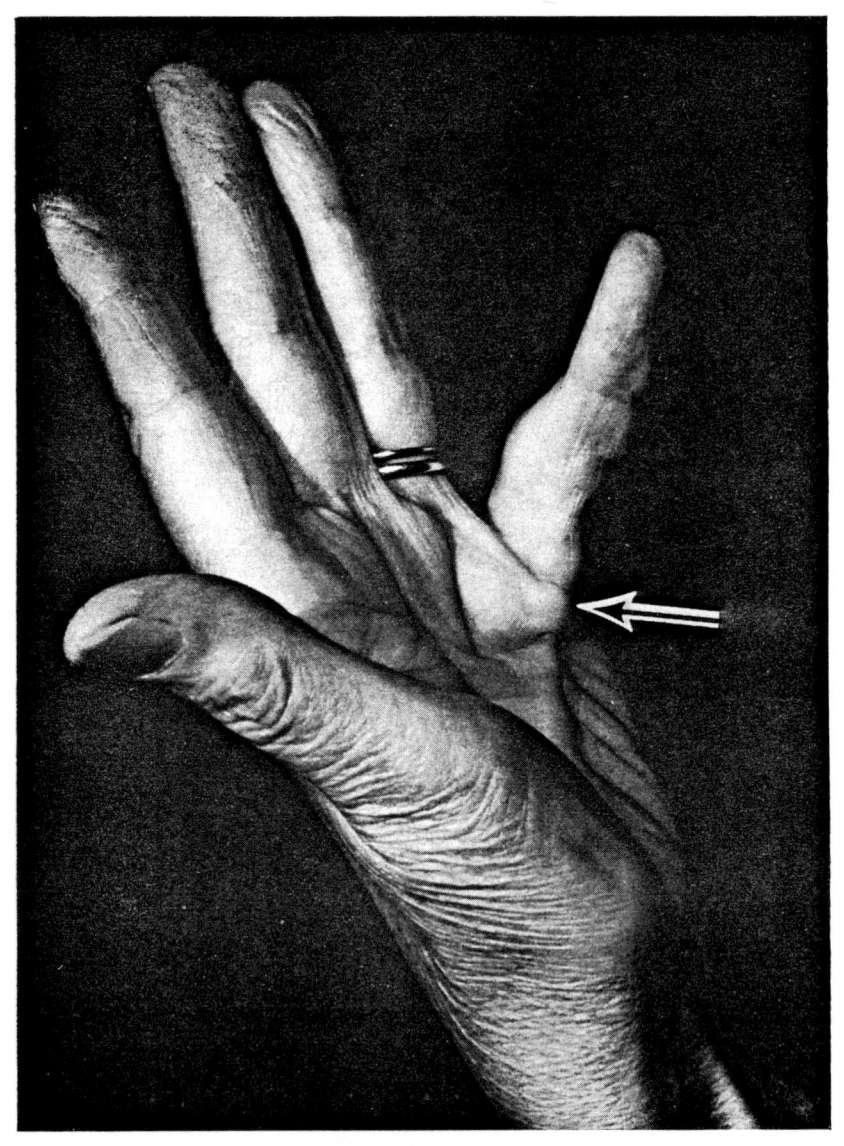

Figure 67. Dupuytren's contracture with flexion deformity of the little finger MP joint. Note the contracted, hypertrophied palmar fascia band (arrow).

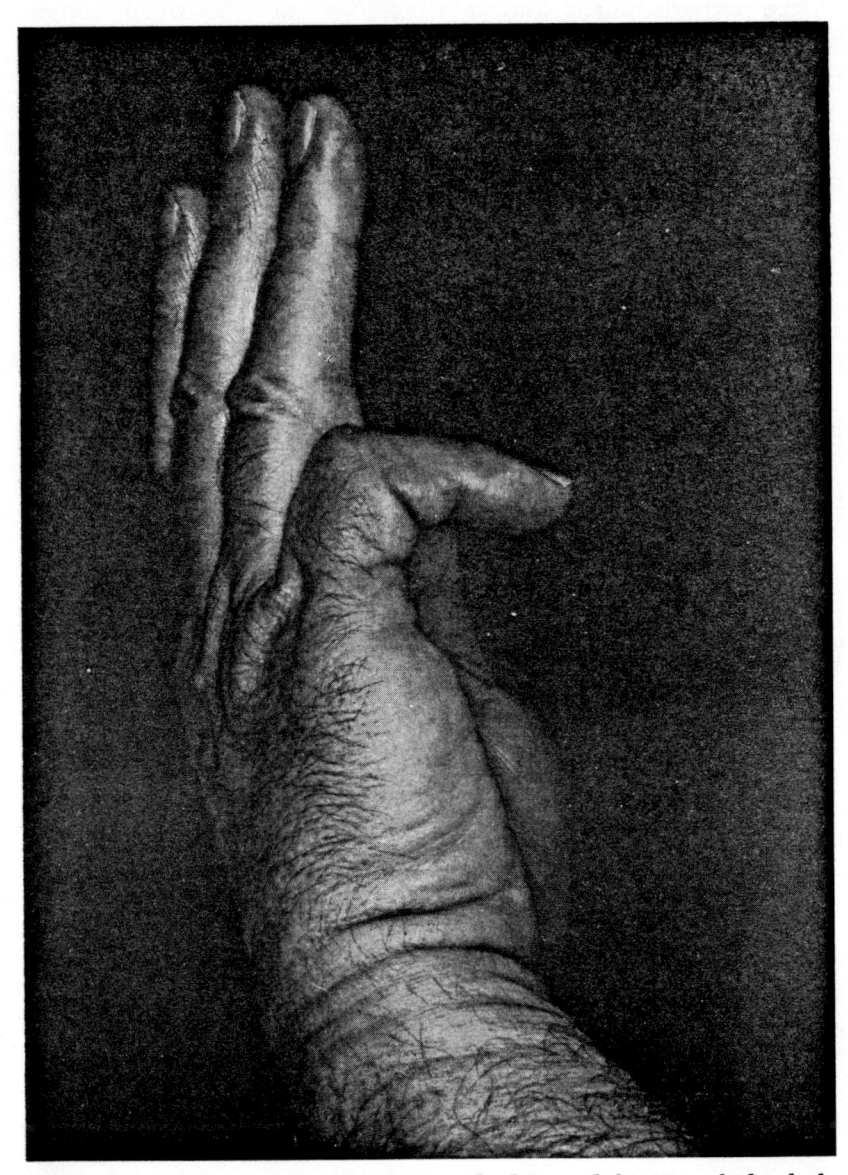

Figure 68. Dupuytren's contracture with flexion deformity of the little finger PIP joint.

The only treatment for Dupuytren's contractures is excision of the hypertrophied and contracted fascial tissue, although surgical intervention is usually not warranted until flexion deformities of the fingers appear. On the other hand, the condition must not be allowed to progress to severe deformities, nor should moderate flexion contractures be allowed to remain for many months, because the development of secondary contractures of the joint capsule, the ligaments, and the neurovascular bundles may make full correction impossible. Flexion deformities of the MP joints are more easily corrected than those of the PIP joints. In extreme cases, amputation of the involved fingers may become necessary.

In general, the optimum time for surgery is when the patient has become unable to straighten one or more fingers, but before the flexion deformity has reached 60° at the MP joint or 30° at the PIP joint. The constitutional response to surgery varies widely among patients subject to Dupuytren's contractures. Some will heal *per primam* and motion of the fingers is rapidly regained. In other cases healing may be complicated by marginal skin necrosis and induration of the hand generally, even including fingers which were not involved in either the disease process or the surgery. Convalescence may be prolonged for many months, and some patients will have significant residual disability. Some will even be worse after surgery than before.

Chapter VII

TUMORS AND MASSES

Carcinoma of the skin occurs not infrequently on the dorsum of the hand, especially in elderly patients. Other primary malignant tumors of the hand are rare. Despite the rarity of malignant tumors, a positive diagnosis must be established of all masses which appear in the hand. The physician should not dismiss a lesion as being benign simply on the basis of statistics.

GANGLION

A mass which appears spontaneously on either the dorsal or volar surface of the wrist is usually a ganglion cyst (Figs. 69 and 70). Occasionally a ganglion may be congenital, but in most cases ganglia develop as a result of focal areas of degeneration in a joint capsule or a tendon sheath. The ganglion may fluctuate in size or enlarge slowly; in some instances the mass does not appear to change in size at all. Some patients complain of pain in the area with use of the hand. The ganglion may be soft or rubbery hard in consistency. It is not adherent to the overlying skin. There is usually no local tenderness. X-rays are negative for bony changes. The diagnosis of a ganglion is proved by aspiration of the material from the cyst. The material may be water-clear or amber in color and may be either fluid or a stiff gel. Ganglia do not undergo malignant degeneration, nor do they produce osteoarthritis. The proper treatment is complete surgical excision, although excision is not necessary unless the mass is a cosmetic problem or is causing significant pain.

There is a high rate of recurrence following surgery if only a simple removal of the prominent sac is performed. The area of degenerated joint capsule or tendon sheath must likewise be

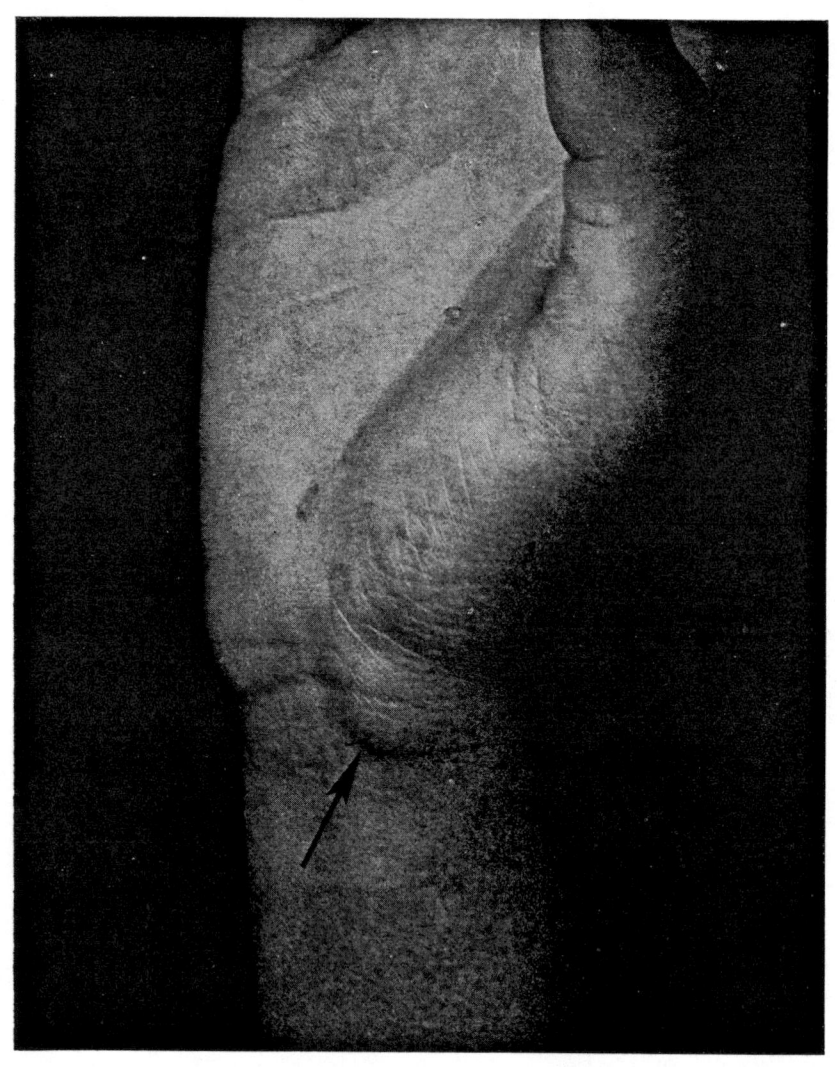

Figure 69. Ganglion of the volar side of the wrist.

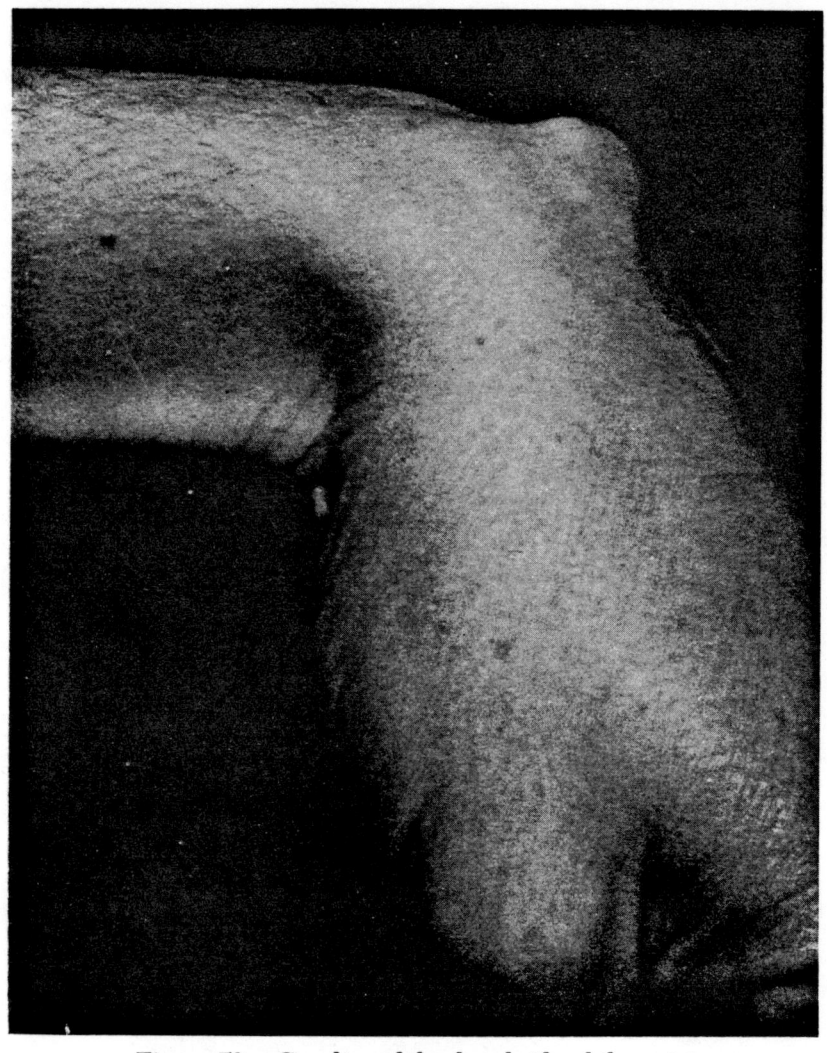

Figure 70. Ganglion of the dorsal side of the wrist.

excised in order to minimize the possibility of recurrence of the ganglion.

XANTHOMA

A xanthoma (benign giant-cell tumor) usually presents as a painless, firm, nonmobile mass in the finger or thumb, particularly on the volar surfaces (Fig. 71). These tumors are benign and tend to grow slowly. They may occur singly, in multiple areas, or in multiple digits. There is no known etiology. Usually they are asymptomatic; however, when present on the tip of a finger or the thumb, the mass may interfer with the function of pinch. The only treatment is surgical removal of the mass.

INCLUSION CYST AND GRANULOMA

A localized mass may result weeks or months after a patient has sustained a penetrating wound. There may be no visible scar, and the minor trauma may not even be remembered. A few cells of the dermis may have been driven beneath the skin by the injury, and an inclusion cyst develops. If, instead, an irritating foreign substance was introduced beneath the skin, a foreign body granuloma will develop. The symptoms may be limited to the presence of a mass, with or without accompanying tenderness. Complete surgical excision of the inclusion cyst or foreign body granuloma is the only effective treatment.

TENDON SHEATH CYST

Cysts of a tendon sheath most often arise in the volar aspect of the hand and involve the flexor tendon sheath at the level of the proximal phalanx. These tumors present as a nontender, semimobile mass less than 5 mm in diameter. The mass does not move with the flexor tendon. These cysts are benign and rarely become large. Unless the mass causes discomfort when the patient grasps objects, no treatment is indicated.

Tumors and Masses 103

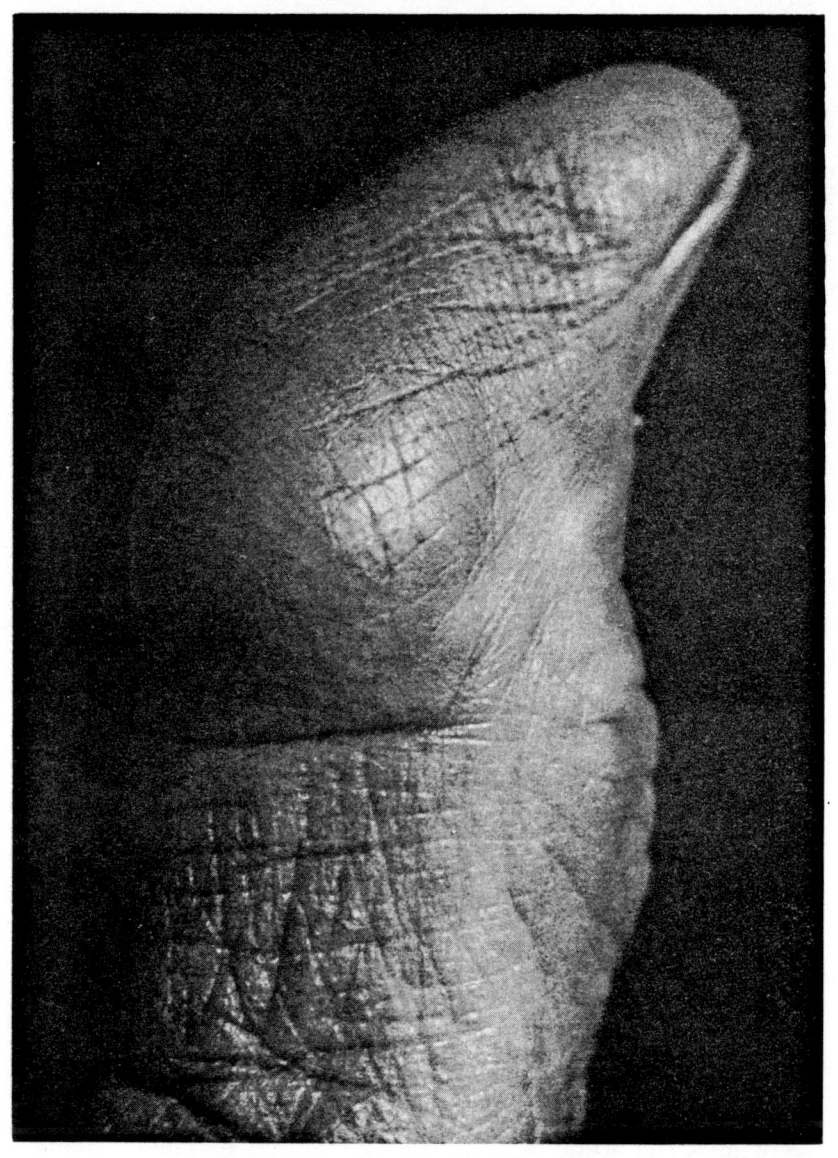

Figure 71. Xanthoma of the thumb.

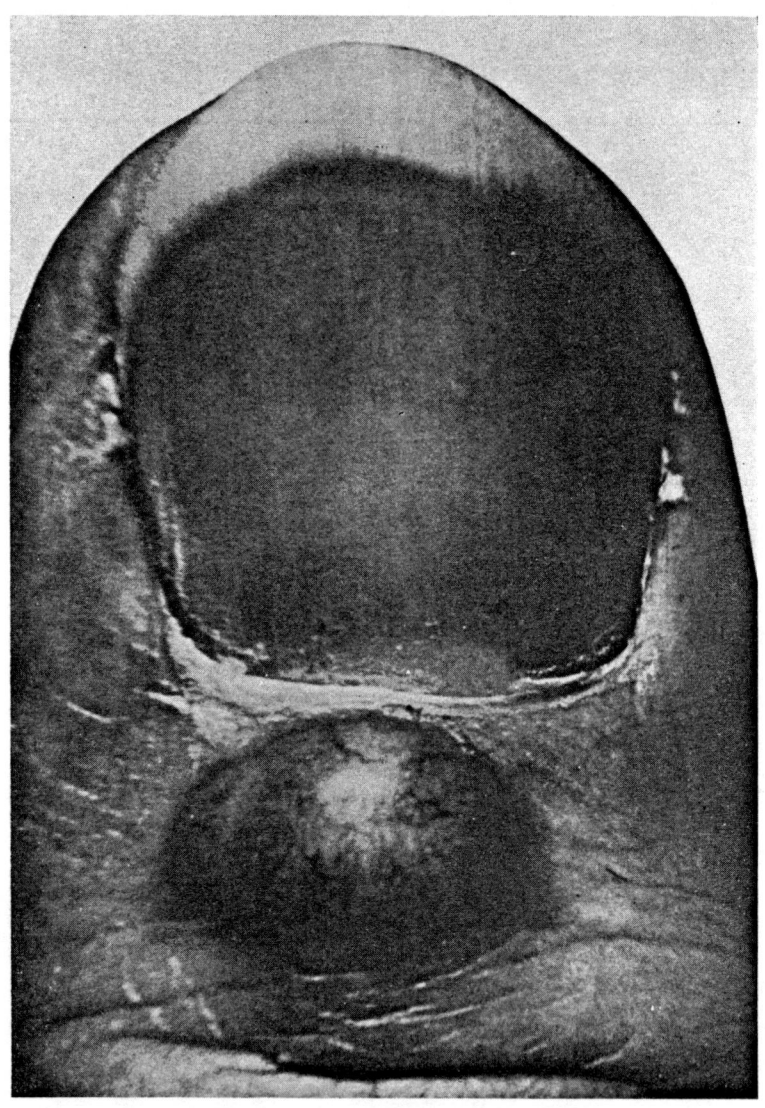

Figure 72. Mucous cyst of the thumb.

MUCOUS CYST

Mucous cysts occur in middle-aged or older patients, usually over the dorsum of the DIP joint. The underlying joint is frequently osteoarthritic, and may communicate with the cyst. The gelatinous content of the cyst may be visible (Fig. 72) as the

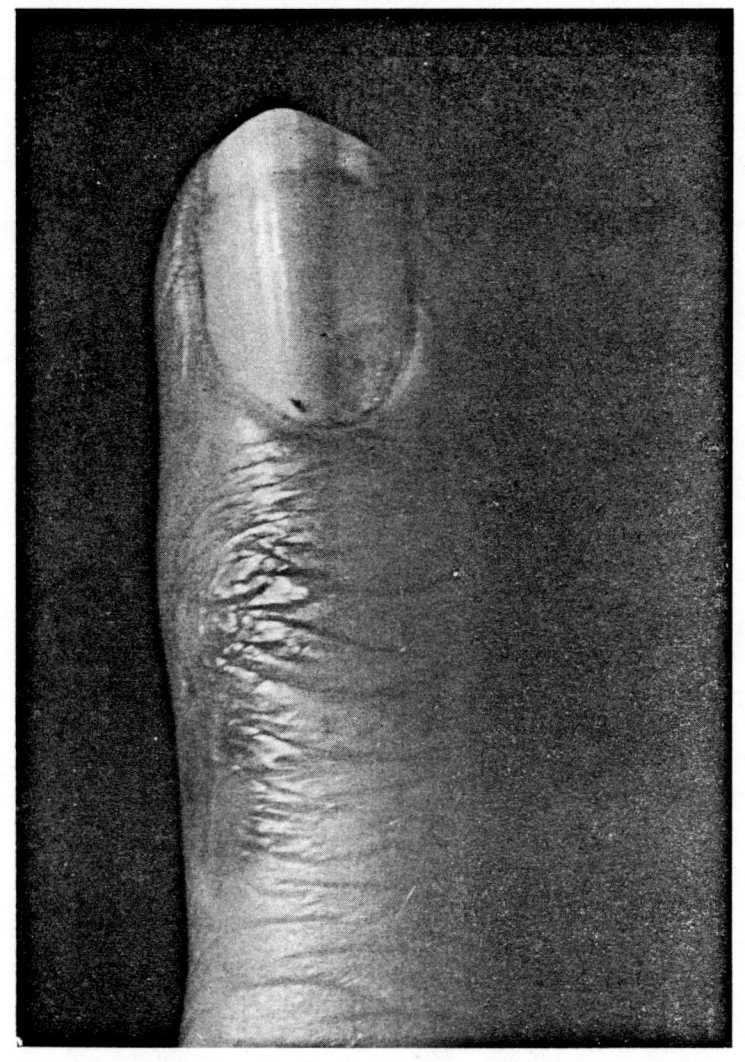

Figure 73. Grooving of a finger nail after removal of a mucous cyst.

cyst thins out the skin. Aspiration of the material rarely cures the condition. The cyst and the overlying attenuated skin must be excised. The defect is then closed by advancing a flap of skin. Some degree of grooving of the nail may result from excision (Fig. 73), and some cysts will recur. These cysts are benign.

106   The Hand Book

CYSTIC BONE LESIONS

Single or multiple osteolytic lesions of the hand are usually either simple bone cysts (Fig. 74) or benign enchondromata (Fig. 75). These conditions are usually diagnosed as incidental findings on x-rays taken for other reasons. The lesions are painful only if the patient sustains a pathologic fracture. No treatment is necessary unless a patient sustains such a fracture or unless the

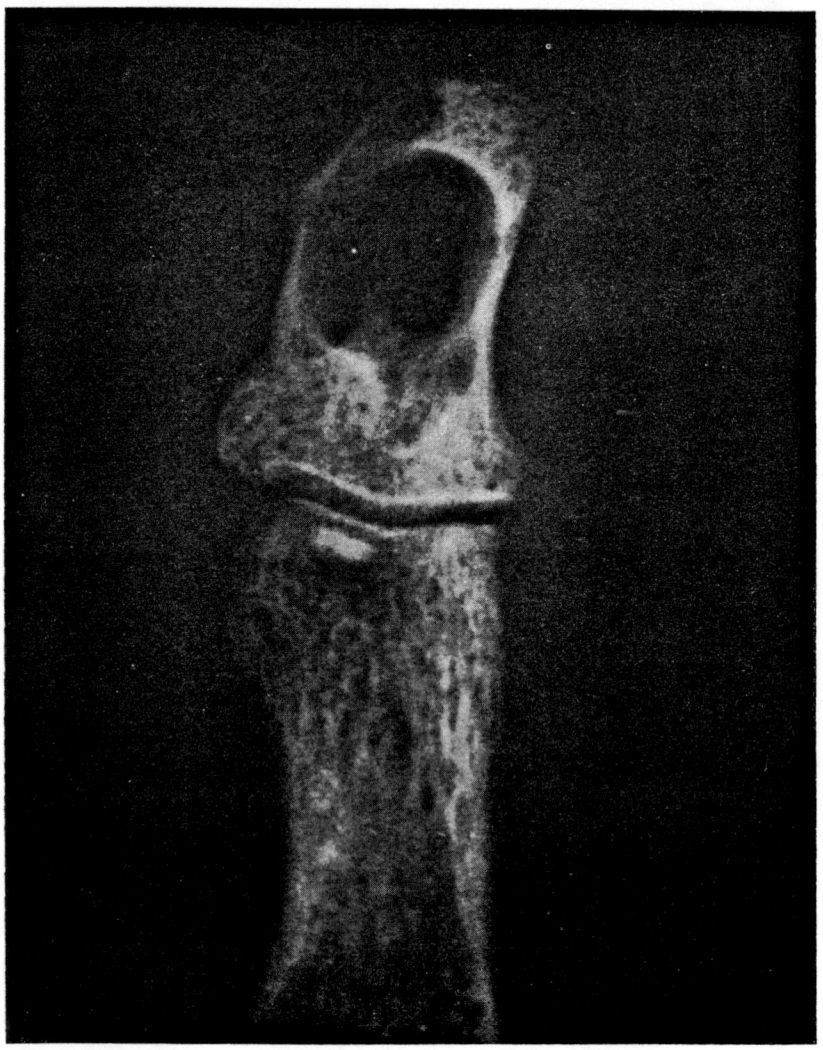

Figure 74.   Bone cyst of the distal phalanx of the thumb.

bone is so thin that there is a high probability of a fracture occurring. The proper operation is to curette the lesion and pack the cavity with bone chips. Normal architecture will be restored by this procedure (Fig. 76).

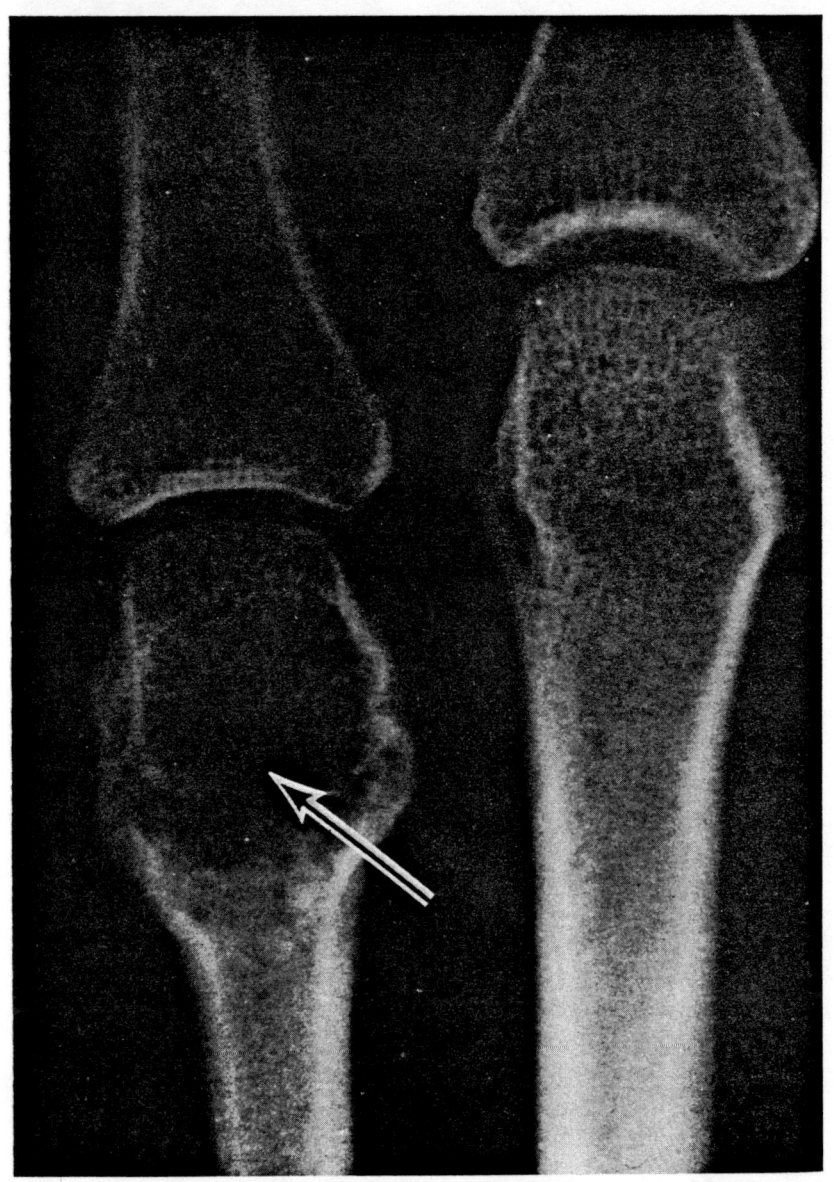

Figure 75. Benign enchondroma of the third metacarpal head.

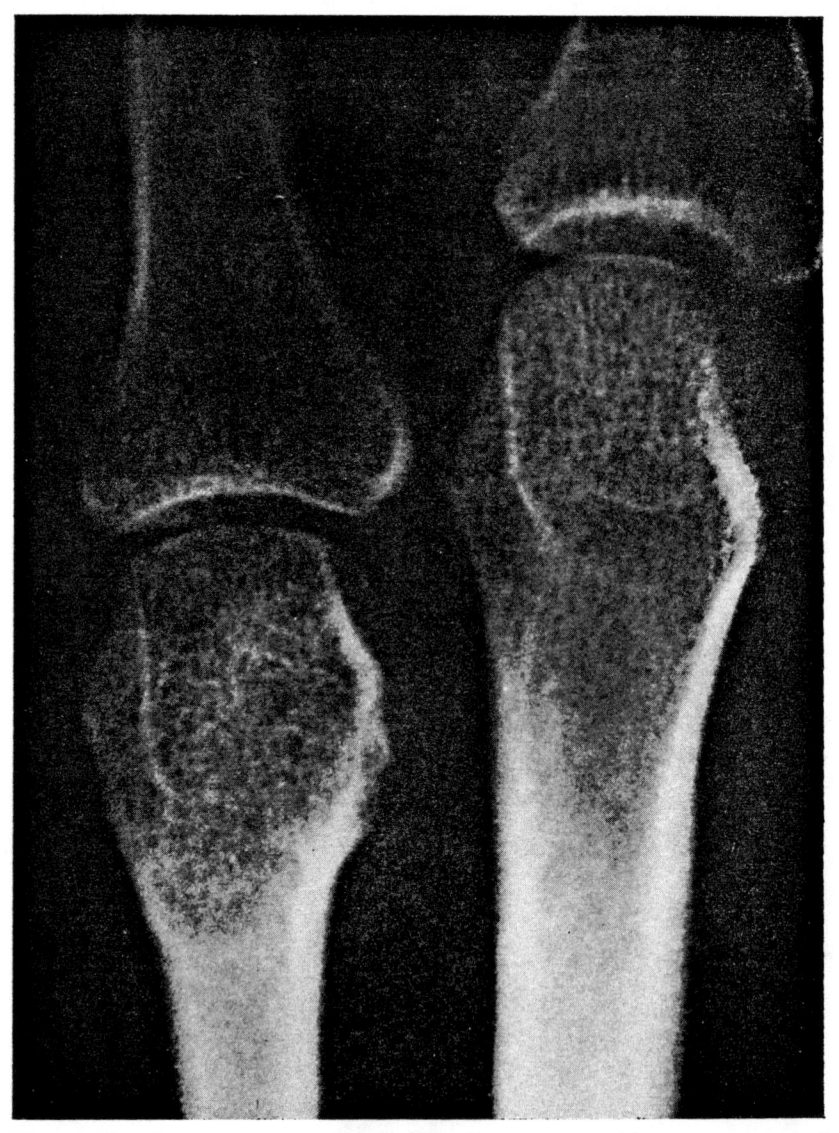

Figure 76. Same patient as Figure 75 after surgery. Note restoration of normal bony architecture one year later.

INDEX

Abductor pollicis longus
  laceration, test for, *14*
  relationship to anatomic snuffbox, 36, *38*
  tenosynovitis, 69
Abscess, *see* Infection
Adductor pollicis, 9, 10
Amputation, 22
  finger, 23
  in Dupuytren's contractures, 98
  ray resection, 23
  thumb, 22
Anesthesia, median nerve, 8
  ulnar nerve, 9
Anesthetics, local, diagnostic use, 7, 19
Antibiotics, 62, 66
Artery, radial, 12
  ulnar, 13, 19, 71
Arthritis, *see* Rheumatoid arthritis, Osteoarthritis
Arthrodesis, definition, 79
  in osteoarthritis, 48, *78*, 79
  in rheumatoid arthritis, 90, *94*
Arthroplasty, definition, 79
  in osteoarthritis, 79
  in rheumatoid, 87, *88*, 89
Atrophy, thenar, 80, *81*
Avulsions, ligament attachment, 57, *58*
  skin, 26
  tendon insertion, *58*, 59

Bite, human, 66
Bone, *see* specific bone by proper name
Bone cyst, 106
Bone graft, 90, *94*, 107
Burns, 27
  chemical, 29
  electrical, 29
  thermal, 27, *30*, *31*, *32*

Calcific tendinitis, 69, *71*
Carcinoma, skin, 99

Carpal tunnel syndrome, 80, *81*, 82
Casts, for Colles' fracture, 33, *34*, *35*, 37
  navicular, *41*
  phalanges, 53
Causalgia, *see* Reflex sympathetic dystrophy
Colles' fracture, *see* Fractures
Contracture, Dupuytren's 95, *96*, 97
  joint, 29, 33, 51, 61
Cyst, bone, 106
  ganglion, 99, *100*, *101*
  inclusion, 102
  mucous, 104, *105*
  tendon sheath, 102

Deformity, mallet, 59
  rotational, of phalanges, 54
Degenerative diseases, 67
DeQuervain's disease, 69
DIP, 5
Dislocations, carpal, 43
  metacarpo-phalangeal, 51, *52*
  navicular, 43, *44*
  in rheumatoid arthritis, 92, *93*
Distal interphalangeal joint, 5
Dorsal interosseous, 10, *12*, 15
Dorsal surface of hand, 3, 5
Dorsiflexion, *see* Extensors, wrist
Dupuytren's contracture, 95, *96*, 97
Dystrophy, reflex sympathetic, 60

Edema, in burns, 29
  in fractures, 33
    Colles', 33
    phalangeal, 51
  in skin avulsion, 26
  prevention of, 33
  stiffness from, 29
  treatment of, 33
Enchondroma, 106, *107*, *108*
Evaluation of lacerations, *see* specific structure

Examination of lacerated hand, see
    specific structure
Extension, see Extensors
Extensors, finger and thumb
    avulsion of insertion, 58, 59
    laceration, 14
    results of repair of, 11
    role in pinch, 68
    rupture in rheumatoid arthritis, 84,
        90
    traumatic, 59, 60
Extensors, wrist, 15
    invasion by rheumatoid arthritis, 84,
        90
    laceration of, 15
    role in pinch, 68
    see specific structure
Extensor, carpi radialis brevis, 15
    carpi radialis longus, 15
    carpi ulnaris, 15
    digiti communis, 15
    indicis proprius, 15, 16
    pollicis brevis, 69
        laceration, test for, 14, 14
    pollicis longus, 15
        laceration, test for, 14, 14
        relationship to anatomic snuffbox,
            36, 38
Extrinsic muscles, definition, 3

Fascia, palmar, 95
Felon, 62, 63
Fibrosis, 29, 33, 51, 61
Finger, abduction, 10, 12
    abduction, 10, 13
    extension, see Extensors, finger
    flexion, see Flexors, finger
    mallet deformity, 59, 60
    names of, 3, 5
Finklestein test, 69, 70
Flexion, dorsiflexion, see Extensors,
        wrist
    finger, see Flexors, finger and thumb
    Thumb, see Flexors, finger and
        thumb
    volar flexion, see Flexors, wrist
    wrist, see Flexors, wrist
Flexor, carpi radialis, 19
    carpi ulnaris, 19

clacific tendinitis, 69, 71
digitorum profundus, 15
    laceration of, test for, 16, 17, 18
digitorum sublimis, 15
    laceration of, test for, 19, 20
digitorum superficialis, see Flexor
        digitorum sublimis
pollicis brevis, 15
pollicis longus, 15
    laceration of, test for, 17, 19
    rupture in rheumatoid arthritis,
        90, 91
Flexors, finger and thumb, 15
    laceration, 15
    results of repair of, 21
    role in pinch, 68
    rupture, in rheumatoid arthritis, 90,
        91
    see specific muscle
Flexors, wrist, 19
    function of in pinch, 68
    laceration of, 19
Foreign body granuloma, 102
Fracture, Boxer's, 48, 49
    casts, 33, 34, 35, 37, 39, 41, 53
    Colles', 33
        median neuropathy in, 80
        edema in, 33
        immobilization, 33
    metacarpal, Bennett's, 43, 46, 47
        neck, 48, 49
        shaft, 48, 50
    navicular, 36, 39, 40, 41, 42
    pathological, 106
    phalanx, 51, 53, 54, 55, 56, 57,
        58
    radius, distal, see Colles' fracture, 33
    scaphoid, see Navicular, 36
    treatment, principals of, 33
Function, test for, see specific structure

Ganglion, 99, 100, 101
Graft, bone, 90, 94, 107, 108
    skin, 26, 29
    tendon, 22
Granuloma, foreign body, 102
Grasp, mechanics of, 23, 67

Index

Hand function, alteration of, after digit amputations, 23
　mechanics, 67
　role of sensation in, 8, 68
High pressure injection injury, 26, *28*, frontispiece

Immobilization, wrist, 35, *37*
　MP joints, 51
　navicular, 39, *41*
　PIP joints, 51
Injection injury, high pressure, 26, *28*, frontispiece
Inclusion cyst, 102
Infection, abcess, collar button, 66
　antibiotics in, 62, 66
　felon, 62, *63*
　human bite, 66
　of palmar spaces, 62, *65*
　of tendon sheath, 62, *64*
　paronychia, 62
　surgery in, 62
　systemic signs in, 63
Injury, acute, 7
　amputations, 22
　avulsions, ligament attachment, *57*, 58
　　skin, 26
　　tendon insertion, 58, *59*
　burns, 27
　high pressure injection, 26, *28*, frontispiece
　to bones and joints, 23
　　nerves, 7
　　tendons, 11
Interosseous, dorsal, function, 10, 15
　test for, *12*
　volar, function, 10, 15
　test for, *13*
Interphalangeal joints, distal (DIP) of fingers, 5
　flexors of, 15
　fractures involving, 55, *58*
　of thumb (IP), 5
　proximal (PIP) of fingers, 5
　flexors, 15
　fractures involving, 55, *56*

Intrinsic muscles, defined, 3
　median innervated, 9
　ulnar innervated, 9

Joints, *see* under specific name or condition

Lacerations, diagnosis, 7
　local anesthesia in, 7, 19
　hand and wrist, dorsal surface, 15
　palmar surface, 7, 15
　radial surface, 12
　volar surface, 15
　nerve, test for, *see* specific nerve
　tendon, test for, *see* specific structure
Ligament, attachment, avulsion of, 57
　MP joints, 51
Lumbricals, 15

Mallet finger, *59*, 60
Masses, *see* Tumors
Median nerve, *see* Nerve, median
Metacarpal fracture, *see* Fracture, metacarpal
Metacarpophalangeal (MP) joint, dislocation, 51, *52*
　flexors of, 15
　fractures involving, 43, *46*, *47*
　subluxation of, 43, *46*, *47*
Mucous cyst, 104, *105*
Muscle, *see* specific muscle
　Extrinsic, definition, 3
　Extensor, *see* Extensors
　Flexor, *see* Flexors
　Intrinsic, definition, 3

Nail, grooving, 105
Navicular, fracture, 36, *39*, *40*, *41*, 42
　subluxation, 43, *44*
Nerve, median, 19
　carpal tunnel syndrome, 80, *81*, *82*
　Colles' fracture, 80
　opposition of thumb, 8, *11*
　paraesthesia, 80
　sensory distribution, *8*
　severance, functional loss from, 8
　test for, 9
　radial, 12
　repair, 7, 11

sensory supply to
  fingers, 8, 10
  palm, 8, 10
  thumb, 8
ulnar, 71
  opposition of thumb, 9
  paraesthesia, 83
  sensory distribution, 9, 10
  severance, functional loss from, 9
  test for, 9
Neuroma, 13

Osteoarthritis, 72
  carpometacarpal joint, 47, 48, 74
  DIP joint, 60, 73, 74
  navicular non-union, 36, 75
  silicone implants for, 79
  thumb, 76, 77, 78
  wrist, 74, 75

Palmar fascia, 95
Palmar flexion, see Flexors, wrist
Palmaris longus, 19
Palmar fibromatosis, see Dupuytren's contractures
Paraesthesias, 80
Paronychia, 62
Phalanx, fracture of, see, Fracture, phalanx
Pinch, mechanics of, 23, 67
PIP joint, 5
Pisiform, 69, 71
Proximal interphalangeal joint, 5
Prosthesis, finger joint, 79, 87

Radial, nerve, see Nerve, radial
  side of hand, 3, 4
Radius, fracture, see Colles' fracture, 33
Radius, styloid process, 75
  relationship to anatomic snuffbox, 36, 38
  tenosynovitis, 69, 70
  DeQuervain's disease, 69, 70
Reflex sympathetic dystrophy, 60
Repair, see specific structure
Rheumatoid arthritis, 84, 85, 86, 87, 88, 89, 91, 93
  arthrodesis, 90, 94

arthroplasty, 87, 88, 89
median neuropathy, 80
prosthesis, 87
silicone implants, 87
synovectomy, 84
tenosynovectomy, 91

Scaphoid, see Navicular
Sensation, loss, in carpal tunnel syndrome, 80
  in median nerve injury, 8
  in radial nerve injury, 12
  in ulnar nerve injury, 9
protective, 68
role of, in hand function, 68
supply to fingers, 8, 8, 10
  palm, 8, 8, 10
  thumb, 8, 8
tactile gnosis, 9
Sheath, tendon, 21
  cyst, 102
  ganglion, 99, 100, 101
  giant cell tumor, 102, 103
  infection, 62, 64
  see Tenosynovitis
Silicone implants, 79, 87
Skin, avulsion, 26
  graft, 26, 29
Snuffbox, anatomic, 36, 38
Stellate ganglion block, 61
Subluxation, Bennett's fracture, 43, 46, 47
  DIP, 59
  navicular, 43, 44
  PIP, 56
  rheumatoid arthritis, 84, 85
  thumb carpometacarpal joint, 43, 46, 47
Sympathetic dystrophy, 60
Sympathectomy, 61
Synovectomy, 84

Tendinitis, calcific, 69, 71
Tendon laceration, 7
  extensor, results of repair of, 11
  flexor, results of repair of, 21
  test for, see specific structure

Tendon rupture, extensor, 60
　flexor pollicis longus, 90, *91*
　mallet finger, *59*, 60
　rheumatoid arthritis, 90, *91*
　traumatic, 60
Tendon sheath, *see* Sheath, tendon
Tendon transfer, 9
Tendon repair, results of, extensors, 11
　flexors, 21
Tenosynovectomy in degenerative disease, 69
　in rheumatoid arthritis, 91
Tenosynovitis, degenerative, 69
　infectious, 62, *64*
　rheumatoid arthritis, 91
Test, Finklestein, 69, *70*
　for function of nerve or tendon, *see* specific structure
Thumb, opposition, 8, *11*
　role in grasp, 23, 67
　sensory nerve supply, 9, *9*
Trigger finger, 72
　thumb, 72
Tumor, 99
　carcinoma, 99
　enchondroma, 106, *107*, *108*
　ganglion, 99, *100*, *101*
　giant cell, 102, *103*
　granuloma, 102
　inclusion cyst, 102
　tendon sheath cyst, 102
　xanthoma, 102, *103*

Ulna, resection in rheumatoid arthritis, 91
Ulnar, artery, *see* Artery, ulnar
　nerve, *see* Nerve, ulnar

Volar flexion, *see* Flexors, wrist
Volar interosseous, 10, *13*, 15
Wrist extension, *see* Extensors, wrist
　flexion, *see* Flexors, wrist

Xanthoma, 102, *103*
X-ray, normal, hand, *6*
　carpometacarpal joint I, *46*
　finger, *56*
　wrist, *45*

RD
559
J63

Johnson, Moulton
Hand book.

DATE DUE			

DRAUGHON'S JUNIOR COLLEGE
709 MALL BOULEVARD
SAVANNAH, GEORGIA 31406